Public goods and decentralization

Tilburg Studies on Economics

10

A study on Mathematical Economics

Public goods and decentralization

The duality approach in the theory of value

P.H.M. Ruys

Research Fellow at the Department of Econometrics, Tilburg School of Economics, Social Sciences and Law

Foreword by

J.J.J. Dalmulder

Professor of Econometrics and Mathematical Economics, Tilburg School of Economics, Social Sciences and Law

1974
Tilburg University Press
The Netherlands

Printed in Belgium

Distributors: Academic Book Services Holland, P.O. Box 66,
Groningen, the Netherlands

ISBN 90 237 2916 1

Foreword

The introduction of public goods into economic theory has recently received substantial attention from economists.

Both the increasing significance of public goods and the deficiencies of the private goods—oriented allocation mechanisms call for a re-orientation of the classical general equilibrium models such as the one designed by Walras exactly a century ago.

Although several approaches to this problem are possible, the duality approach seems to be very fruitful. In this study, Dr. Ruys applies and develops the mathematical theory of polarity to gain a deeper insight into the economic theory of value. He not only uses the duality approach to simplify formal proofs related to public goods, but he also gives an interpretation to the application of polarity operations which goes far, beyond its formal significance.

This study is in a way closely related to the publication 'Axiomatic choice models' by Dr. H.N. Weddepohl, which is volume 3 of the Tilburg Studies on Economics.

The book consists of two parts which may be read independently; the first on economics and the second on mathematics. The mathematical part consists of three chapters; the first (chapter 8) giving an introduction on notions and notations used, the other two developing the theory of duality for sets and multifunctions. In the economic part, the theory of value is developed for producers and consumers (chapters 3 and 4) and related to (local) public goods. Several approaches to the equilibrium problem are described in chapter 5, including the introduction of a two-level price equilibrium.

In chapter 6 institutional and procedural implications are analyzed and chapter 7, finally, enlarges upon the theory developed.

J.J.J. Dalmulder

Acknowledgements

Many people have, of course, a direct or indirect influence on the formation and composition of ideas motivating somebody to carry out a research project. Below I mention some of the people influencing my work, although it should be stressed here that if gratitude were to be adequately expressed this list would become much longer.

First of all, I would like to thank Professor Jan J.J. Dalmulder for providing a stimulating and creative environment in which to work: his experience in academic life and his view on 'la condition humaine', have both inspired me and shown me the relativity of results and deficiencies. I owe also much to Dr. Claus N. Weddepohl, with whom I have worked closely and who introduced me to the world of convexities and dualities. In particular, I learned from him to consider mathematical scrutiny as being a moral element in science.

Professor Lionel W. McKenzie taught me Walras' significance for economic theory, which has had a great impact on my subsequent research. A similar result was obtained by some stimulating remarks made by Dr. Benjamin Breek, some years ago.

Professor Jacques H. Drèze and Dr. Leo R.J. Westermann have made many valuable comments on the preliminary draft of this study, and the book has greatly benefited from their suggestions.

The form of the book has been substantially improved by the drawings made by Mr. Ad M. van Helfteren, and the editing done by Mrs. Carolyne Brand-Maher. I would also like to thank the following institutions for the support extended me in this study: the Katholieke Hogeschool in Tilburg, the University of Rochester, N.Y., and the International Economic Association.

The burden of writing this book was carried as much by my wife, Ireen, and our children, as by myself: it is therefore really our work. For a similar reason this book belongs to my parents, and to the friends and relatives with whom we shared many goods long before the economic concept of a public good entered our lifes.

Contents

I. Economics

1. Introduction

1.1. ECONOMIC THEORY AND ECONOMIC ENVIRONMENT

In an essay on the progress of economic thought, Joan Robinson calls the attention of her reader to the satire by Bernard Mandeville, written about 1700, called 'The Fable of the Bees'. The bees were prodigal, luxurious and vain, and all these vices made their society flourish, fully employed and wealthy. One day, she notes, the bees were 'smitten with virtue and began to lead a sober life, eschewing pomp and pride, and adopting frugal, modest ways. The result was a dreadful slump' (1962, p. 14). The paradox that private vices can be public benefits has intrigued many economists, from Adam Smith on. This inspires Mrs. Robinson to describe the task of the economist as follows: 'It is the business of the economists, not to tell us what to do, but to show why what we are doing anyway is in accord with proper principles' (p. 21). This statement indicates quite clearly the intertwinement of economic theory and ideological environment. Any economic system, according to Mrs. Robinson, requires a set of rules, an ideology to justify them, and a conscience in the individual which makes him strive to conform to them.

It is not only the ideological environment that has a relation to economic theory, but also the technological environment in the widest sense. The transformation of an economic system is determined by the changing environment, but at the same time the environment is influenced by the prevailing economic system and theory. A well-known example is provided by the capitalist system in which the allocation of resources is completely determined by the competitive market. The working of the capitalist system with its individual-oriented ideology of progress has stimulated the development of economies, but has also resulted in the accumulation of the forces of production into private hands. At the same time it has changed the production conditions by developing new processes with indivisibilities, increasing returns to scale and external effects. Gradually, the increased productive forces no longer fitted into the system which had caused them to develop. Market failure, imperfect competition, unemployment and inequity undermined the foundations of the capitalist system, and new rules became necessary.

New economic systems have been developed, mainly under the pressure of political forces, and observed and influenced by political economists who (also!) have to analyze both the prevailing systems and theories, and the environment, in order to adapt the economic system to the changing environment. Gradually, however, the analysis of allocation mechanisms has received more attention, which implies a more active attitude on the part of political economists towards the design of an economic system.

A third type of environment on which economic theory depends is its language. It is very difficult to adequately describe and analyze complex economic phenomena in a verbal way. Therefore mathematical and other tools of analysis are used in economic models. Although these formal languages also have their deficiencies in the description of economic phenomena, they do make it progressively more possible to describe social features or systems in a mathematical model, or conceptual system. Some examples are: optimizing behavior (programming), conflicts (game theory), (im-) perfect competition (measure theory), uncertainty and risk (probability theory), general equilibrium (topology), economic policy (control theory), and economic organization (system theory).

Koopmans (1957) sees economic theory as a sequence of models, derived from the interaction of observation and reasoning. This dialogue has been intensified only during the last few decades. The postulational approach (see Nauta, 1970) holds many advantages, both for scientific efficiency and for the communication between sciences and scientists, and has therefore gained in momentum.

For many models of aggregate economic behavior (or macro-economic models) this approach is based upon and verified with empirical data. Unfortunately, this cannot yet be said of micro-economic models, in which assumptions are made concerning the behaviors of individual agents.

The present study does not belong to the category of positive or descriptive models of an economy which can be empirically verified, but to the category of normative models based upon postulates and giving recommendations for an economy where these postulates are valid or acceptable. This branch of economics is usually called welfare economics. On the other hand, the postulates and definitions to be chosen must describe – as closely as possible – the essential characteristics of agents and features in an economy. The over all performance of a formalized economy with purely economic agents and pure concepts can be very informative for people who have to design and improve economic systems in real life. This postulational approach, called pure economics by Walras exactly a century ago, has an operational value only if all relevant features in real life are continuously translated in new and better postulates and definitions.

The present study is concerned with the introduction of public goods in a decentralized economic system. In the case of private goods it is assumed that the total quantity of these commodities is distributed over the individual agents, who have exclusive control of their bundles and are not affected by the bundles available to other agents. The incentive to expand the quantities at the disposal of an agent, causes a conflict of interests between individual agents which can be reconciled in some optimal way through the competitive market mechanism, based on the parametric function of a price system (chapter 2). Public goods cannot be exchanged on a market and are available equally to all agents in the economy. Therefore, the conflict of interests between individual agents does not involve the quantities at the disposal of an agent, but the composition of the bundle of public goods at the disposal of all agents together i.e. on the valuation of the bundle. To reconcile the conflicting interests about prices among individual agents, mechanisms are described and designed (in chapters 5 and 6) which equalize the benefit of every public good for all agents together and the cost of the public good for all agents together.

In order to do so, much attention must be paid to the theory of value and its mathematical counterpart, the theory of duality. The correspondence between the price structure and the technology and preferences are analyzed in chapters 3 and 4. The mathematical tools are described and developed in part II of this book, (see also section 1.4). In chapter 7, finally, the economic and methodological background of this study is described.

1.2. PRIVATE AND PUBLIC GOODS

A commodity or service is called an economic good if it is desired by a number of individuals in the economy for some characteristic, and if it is scarce, i.e. available only by offering some other economic goods in exchange. Economic goods are sometimes consumed by one individual, sometimes by a number of individuals simultaneously. The consumption of an economic good sometimes has effects on non-consuming agents; sometimes there are no such external effects. These properties of economic goods are decisive for the way in which they can be treated in economic theory.

This study deals with the allocation of resources and producible commodities among the individuals or agents in an economy. Until recently, the theory of optimal allocation was mainly concerned with economic goods which could be handled on a market. A market is particularly apt for economic goods called *private goods*, which are by definition exclusively consumed by one agent so that its consumption has no external effects on other agents (see also section 4.2).

Given an income distribution, or a distribution of resources, the market mechanism invites each consumer to demand an optimal element according to his individual taste and wealth. The information about demand and supply of other agents in the market is transmitted through a price vector, uniform for all agents. This uniform price for private goods is closely related to the fact that individual demands for private goods can be added to give total demand in the market.

Private goods in the strict sense are in fact an exception in an economy rather than a rule. Many goods have *external effects* in consumption, because they enter into two or more persons' preference functions simultaneously. For example, for many people, food, clothes, or a gardiner's service are private goods in the strict sense. But these goods, and many other may have external effects in consumption: e.g. a dinner in a restaurant, a dress, or a garden.

If an economic good is consumed by all agents in the economy in such a way that no agent is aware of this consumption by others ('pure externability'), then it is called a *public good.* This definition is made more precise in section 4.2. Examples are: national defense for a country, clean air for a region, public facilities for a city. If the economy is given and the set of agents consuming collectively a public good is a strict subset of the agents in the economy, then this good is called a *local public good,* such as a soccer match, a dike and control of air pollution.

Again, the concept of a (local) public good is a formal description which will not be completely accurate in most cases. This definition does, however, come very close to many actual economic goods and is also operational in the sense that it serves as a corner stone for a theory of allocation of economic goods in an economy.

There are operations which transform some economic goods from public goods to private goods or vice versa, for instance, the property of non-excludability (required for public goods) can be converted into excludability (required for private goods). A public park, e.g., can be converted into a private garden, and a private collection into a public collection. The ownership of an economic good does not necessarily determine the economic character (private or public) of the commodity. On the other hand, private ownership of public goods (or vice versa) often makes it difficult to obtain an optimal allocation (see chapter 6). It is virtually impossible to transform the character of some economic goods, such as a radio broadcast, or a traffic regulating police officer. Obviously, excludability is difficult to achieve in these cases.

Public goods can also lose their character when consumers become aware of their simultaneous consumption. The enjoyment of consumption of a good is decreased or increased by another individual's consumption of the same good. A well know example is congestion, e.g. of traffic.

Thus, two features determine the character of an economic good: the technical or objective criterion of excludability and the individual or subjective criterion of external effects (see table 1.2.1):

Table 1.2.1. Types of economic goods

economic good:	objective use:	subjective use:
private	exclusive	no external effects
with externalities	(non-) exclusive	external effects
public	collective	pure external effects

Many public goods have a disutility for consumers, such as pollution or inflation. In those cases, the positive definition would be the removal or the prevention of harmful situations. Free disposal of (consumed) commodities also belongs to this category, in my opinion, if it is indeed an economic good (and thus not 'free' or without social cost). Other definitions and interpretations of public goods are given in section 4.3.

Public goods have, of course, always been present in an economy. Two – independent – problems in connection with public goods have to be solved by the community:

1. Which bundle of public goods has to be provided for the agents in the economy?
2. How should it be produced or financed?

The early economists concentrated their attention mainly on the second question: 'expenditure somehow got decided and had to be paid for' (Samuelson, 1966, p. 99). This tradition is set forth in the theory of public finance. Since the neo-classical writers of the 19th century, the problem of optimal taxation has been considered from two distinct points of view: the benefit approach and the ability-to-pay approach (see Musgrave, 1959). The first approach is individualistic, since it questions the right of a goverment to apply other norms for taxation than the benefit the individual agent receives through the supply of public goods. According to Samuelson (1969), this approach is also legalistic, because the acquired rights of the individual agents are considered inviolable, and all should be treated similarly if taxes must be levied, whatever the individual's initial position may be.

The ability-to-pay approach contends that those with greater ability to pay, should pay more in order to get a greater welfare for all. Both approaches leave problems unsolved: the second by assuming that a social welfare function exists and a socially optimal distribution can be determined, the first by assuming also that both the existing and final allocation are socially optimal.

What is socially optimal? The answer to question (1) above also contains the answer to this question. Presumably, it has long been thought that no economic considerations are involved in the assumption about the existence of a social welfare function, and that either an autocratic or democratic political mechanism can lead to a social preference ordering. However, it has been shown by Arrow (1951) that, given some minimal conditions, there is no rule for deriving a social preference ordering from individual preference orderings. This was already noted by Condorcet in 1785 in connection with the majority rule with voting (see also Sen, 1970). Therefore, if one is interested in a decentralized system in which individual preferences somehow determine social preferences, one cannot assume that the government is acquainted with a social preference ordering through a political mechanism.

An interesting way to circumvent this problem has been suggested by Schumpeter (1943) and analyzed by Downs (1957). They assume that governments and political parties competitively maximize consumers' votes in the economy. When consumers vote in their own interests, the government decision is – by definition – optimal. In this approach, however, economic theory cannot offer any criterion to compare different allocations. Which outcome is socially optimal depends on the performance of the political mechanism, and it remains socially optimal even if economists can point to unnecessary losses or to costless improvements with respect to the allocation, for example. In this context, Stevers (1967) has noticed that government policy under vote maximization would be characterized by: a high degree of continuity, a narrow nationalistic promotion of interests, and insufficient attention to long term developments.

The normative approach to public expenditure was introduced by authors as Dupuit (1844), Pantaleoni (1883), Sax (1887), Wicksell (1896), Barone (1912) and Lindahl (1919), whose papers are translated in Musgrave and Peacock (1958). It was Lindahl's approach that inspired Samuelson to formulate necessary and sufficient conditions for an optimum in an economy with private and public goods (see Samuelson 1954, 1955 and 1958). Samuelson introduced individual prices for public goods, and expressed these prices in terms of a private good: the numéraire. His quasi-equilibrium (later called Lindahl-equilibrium) solved both the problem of providing public goods and the problem of financing them, in such a way that a synthesis was obtained between the benefit approach and the ability-to-pay approach.

This solution, however, is purely theoretical as Samuelson himself has noted. Since consumers have to pay a fee according to the individual benefit they receive from a commodity which will be supplied anyhow, it is unrealistic to assume that they will reveal their true preferences. Several authors, viz. Drèze and de la Vallée Poussin (1971) and Malinvaud (1972) have designed procedures to overcome this problem of incentives.

Table 1.2.2. Government expenditures, by economic category – Percentage of gross national product: 1957 and 1967.

	Germany		*France*		*Italy*		*Netherlands*		*Belgium*	
	1957	*1967*	*1957*	*1967*	*1957*	*1967*	*1957*	*1967*	*1957*	*1967*
gross consumption	14.3	16.7	14.6	13.7	11.2	12.8	14.6	15.5	9.9	12.7
direct investment	3.0	4.7	2.3	3.5	1.8	2.3	4.6	5.7	2.5	4.3
income transfers	16.7	18.5	17.7	21.4	13.8	18.0	12.3	17.9	12.5	16.9
assistence, interest[1]	4.3	3.0	4.1	2.7	3.4	3.1	7.9	8.9	3.8	5.4
total	38.3	42.9	38.7	41.3	30.2	36.2	39.4	48.0	28.7	39.3

[1]. Interest paid, capital transfers, loans, advances, and participations.
Source: E.E.C. (1970, tables A1 and A2).

In this study, an attempt is made to solve the problems related to public goods by separating the financing of public goods from the provision of public goods. The individual prices introduced by Samuelson are given a quite different interpretation, and – like votes – are applicable only on the level of public goods. The formal optimality conditions are, of course, the same. The financing problem is solved by a taxation system which is independent of the public goods to be provided, similar to Bowen's 1943 suggestion. This approach is made possible by application of the duality theory, and has resulted in a concrete allocation mechanism.

Table 1.2.3. Government expenditures, by major function – Percent distribution: 1957 and 1966.

	Germany		*France*		*Italy*		*Netherlands*		*Belgium*	
	1957	*1966*	*1957*	*1966*	*1957*	*1966*	*1957*	*1966*	*1957*	*1966*
1. general government	5.8	5.4	3.9	3.7	9.3	6.7	5.2	7.9	5.0	4.5
2. the judiciary and police	3.1	2.8	3.0	2.5	4.0	5.1′	2.9	2.9	3.4	2.8
3. national defense	8.8	9.9	18.0	10.4	9.7	7.9	14.3	8.3	12.7	7.6
4. foreign relations	0.6	1.1	5.1	3.3	0.6	0.4	2.1	1.5	0.9	2.3
5. transportation, traffic	6.2	7.6	7.8	8.0	8.6	7.6	9.8	9.5	11.8	12.7
6. industry and commerce	3.9	2.7	6.8	6.8	1.3	2.4	3.3	3.8	3.2	2.8
7. agriculture	5.5	4.0	2.8	3.3	5.4	4.5	5.6	3.5	1.3	2.3
8. education, culture	7.8	10.1	8.3	13.1	9.7	13.9	12.6	16.4	12.7	14.9
9. welfare	33.0	35.3	30.7	36.8	30.9	38.1	25.2	33.0	32.9	34.9
10. health	4.0	5.6	0.8	1.8	4.5	3.1	2.6	3.3	1.7	1.7
11. housing	5.2	3.6	3.5	4.4	1.4	1.0	8.7	7.8	1.2	1.1
12. disasters and war payment	11.9	7.4	5.5	2.6	5.3	2.9	1.4	0.2	5.8	2.7
13. undistributed	4.2	4.6	3.8	3.3	9.3	6.7	6.3	1.9	7.4	9.7
total	100	100	100	100	100	100	100	100	100	100

Source: E.E.C. (1970, table A10).

Table 1.2.4. Government expenditures, by level of government[1] *– Percent distribution: 1957 and 1966.*

	Germany		*France*		*Italy*		*Netherlands*		*Belgium*	
	1957	*1966*	*1957*	*1966*	*1957*	*1966*	*1957*	*1966*	*1957*	*1966*
the state	54	50	59	50	52	45	42	30	51	52
lower public bodies	18	21	14	16	18	18	41	44	20	15
social insurance	28	29	27	34	30	36	17	26	29	33
total	100	100	100	100	100	100	100	100	100	100

1. Transfers between public bodies and redemptions excluded.
Source: E.E.C. (1970, table B1).

The increasing relevance of public goods can be verified – if necessary – by the preceeding tables, indicating the relative importance and composition of government expenditures for several E.E.C. countries.

1.3. THE DUALITY APPROACH

An often applied device for solving mathematical problems is to translate the concepts in which the problem is described, into other concepts which simplify the solution of the problem. Examples are the Laplace transformation (from time- to frequency-domain), and the transformation from the primal to the dual problem in linear programming. Such a transformation is in itself valuable, as it can give a deeper insight in the problem under consideration.

For both reasons, the duality transformation is applied in this study. A third reason for relevance of the duality approach in the social sciences is, that both representations of the problem may be influenced independently (see chapter 7). This is not assumed to be true in the main part of this study.

Duality phenomena are also well known in economic theory: maximizing profits and minimizing costs, for example, and: maximizing the value of output and minimizing the value of shadow prices in a programming problem. It is essential that there be a common concept connecting both problems.

The mathematical duality concept which is basic to the dual concepts used in this study, is defined in section 9.1 as the set of all linear functions which map a vector of a finite-dimensional euclidean space (the primal space) into a real number. Thus, the real line connects in a way the primal space and the dual space.

Since quantities of an economic good are assumed to be represented by the real line and quantities of n commodities by the real euclidean n-space, this space is called the *commodity space* or *quantity space*. The quantity space is considered to be the primal space. The meaning of the dual space in economic theory depends completely on the connotation given to the real line common to the primal system and the dual system. This connotation can be for example: weight of gold, amount of money, or hours of homogenous labor per year. Since these concepts determine the meaning of the elements in the dual space, such a concept is called the *standard* or *denominator* of the dual systems.

The Walrasian economic theory and most contemporary western economic theories assume a single standard: mostly money. The value of a commodity in these theories is indicated by an amount of money per unit quantity of the commodity, i.e. an element of the dual space. Values and prices are identical in these theories.

The Marxian economic theory knows two independent standards: money and (homogenous) labor. The value of a commodity is expressed in terms of labor, the price in terms of money. The 'profit' of producers in terms of labor values (which are determined by the amount of labor socially necessary to reproduce a unit of the commodity considered) indicates the surplus value of production. Since the labor values need not be proportional to the money prices, the distribution of profits is not proportional to the distribution of surplus values.

It is possible, of course, to introduce other standards of measurement for economic values, such as the amount of counters, or 'votes' per unit quantity of a commodity. In this study (chapter 5) such a standard will be proposed for public goods, not independent of the money standard for private goods but carefully linked with it. Probably, the 'vote' standard offers better opportunities of determining a socially desired output of public goods than the money or labor standards proposed thus far. Further, since the methodology applied is independent of the standard chosen, proportional solutions exist for all standards.

The methodology applied in this study to derive dual concepts is based on fixing the standard on an arbitrary real number – say 1 – and determining the prices (or values) in the dual space which assign the (mathematical) value of at least, or at most, 1 to the vectors in the primal space. This was first done in economic theory by Hotelling (1932) and Roy (1942) who deduced that a consumer's utility function with arguments in the quanty space can be replaced by an 'indirect' utility function in the price space only if the prices are market prices and the consumer's income is fixed. (See section 4.1.)

Duality in consumption theory has been studied independently by Milleron (1968) and Weddepohl (1970). Their findings were extended to production theory by Ruys (1971a), to aggregation of sets by Weddepohl (1972), to dual correspondences (or polar multifunctions) by Weddepohl (1973a, b), and to adjoint multifunctions by Ruys (1974). This adjoint multifunction is a generalization of the adjoint of a convex polyhedral process, defined by Rockafellar (1972). (See chapters 9 and 10.)

Shephard (1953) has shown that in many cases production functions can be derived from cost functions, and the converse relation was shown by Uzawa (1964). Shephard (1970) has generalized these results to multifunctions and derived a dual relation between cost structures (c.q. output revenue structures) in the price space, and production-input structures (c.q. output structures) in the quantity space. (See chapter 3.)

The duality approach in equilibrium theory was first applied by Milleron (1969) to public goods. This was also done – independently – by Ruys (1970). Weddepohl (1972, 1973) has applied this approach to a private goods economy, in order to simplify proof of existence given by Debreu (1962).

One of the main mathematical generalizations made possible by the duality approach is that the requirement of differentiability of functions is no longer necessary. In fact, an element of the dual space can be considered as a generalization of the concept of a differential in calculus. It also allow for the application of all tools developed in set theory. For example, the necessary and sufficient conditions of Pareto optimality given by Samuelson (1954) can be generalized and reformulated in terms of separating hyperplanes (see Fabre, 1969, and chapter 9). Finally, since the behavior of concepts at the origin determines the behavior of concepts in the dual space at infinity, compactness arguments can mostly be applied in either space.

The duality concepts and tools mentioned above are related to the dual structures in programming problems, discovered by Kantorovich (1939) and Dantzig (1948). A relation between these concepts is given in section 10.5. In the present study, the time-dimension is not explicitly treated.

A recent contribution to the analysis of dual structures in linear programming problems over an infinite horizon has been made by Evers (1973).

1.4. DECENTRALIZED ALLOCATION MECHANISMS

A theory of value of commodities depends on the technology of producers and the preferences of consumers, both individually and socially, given the standard in which values are expressed. One problem is to show the interdependence and to formulate criteria for equilibrium and optimality of an allocation (see chapter 2).

Another problem is to design an organization through which the technology of producers and the preferences of consumers can be communicated, so that optimal decisions about allocation can be made. One of the oldest mechanisms developed for private goods is the *market*. Let us cell an agent's set of alternative actions in a given situation his *choice set*. The individual's choice set in a market is then determined by the individual's resources and the prevailing prices of the marketable commodities. Usually, the choice sets determined through a market mechanism are large enough to express the producer's technology and the consumer's preferences, while the prices contain enough information to equalize supply and demand. It has been shown that, if all consumers and producers choose a maximal element from their choice set, an allocation is obtained which is Pareto optimal (see chapter 2).

The market mechanism has some remarkable properties, such as:

a. It is decentralized with regard to decisions.
b. It is decentralized with regard to information about local circumstances.
c. It finances the allocation.
d. It is incentive compatible to a large degree (see section 6.1).
e. It is adaptable to some forms of change.
f. It is locally converging to an equilibrium under some restrictions.

These properties (defined in chapter 6) were mentioned in a discussion between economists in the first half of this century about allocation mechanisms in a socialist economy. This 'socialist controversy' drew attention to the design of allocation procedures for the first time.

It is well known that the environment in the 19th century changed to such an extent that the private ownership economy with the competitive market mechanism as only rule of allocation no longer functioned satisfactorily. Marxian analysis pointed out that a contradiction existed between productive relations (economic organization) and productive forces or capacities of individuals. Many productive capacities had outgrown the individual right of disposal and had become in fact social capacities. The socialist economists therefore advocated abolition of private ownership of production factors. Although they emphasized the importance of planning in a socialist system, concrete proposals to run a socialist economy were hardly made at the time.

The most accepted point of view was a centralized economic organization. The conclusion of the socialist controversy was that it is possible to design a mechanism which is formally similar to a market mechanism, but which is very different in other aspects.

The analysis of allocation procedures has only recently been introduced. Hurwicz (1959) has proposed some criteria for informational decentralization. A different approach, based on uncertainty and limited information has inspired J. Marschak (1955) and Marschak and Radner (1972) to develop a 'theory of teams' for organizations with a common goal. Several authors have studied planning procedures in a macro economic context (see Tinbergen, 1964, and Ellman, 1971).

Malinvaud (1968) has defined some criteria or properties for planning procedures in a decentralized economy, and has analyzed procedures for several economic environments. In subsequent papers published by him and other authors, this environment also included public goods (see sections 6.3 and 6.4).

The presence of public goods has important consequences for the design of decision procedures. First of all, at least one decision has to be made centrally. The finance of the provision of public goods and the true revelation of preferences and production technologies also become problems. It is therefore difficult to develop allocation mechanisms for public goods, if some attractive properties of allocation mechanisms for private goods are also to be realized.

The solution proposed in this study is that each agent be given a choice set of priorities for public goods, comparable with the choice set, or budget set, of private goods. Through these choice sets, the individual agents can reveal their own priorities and preferences, and the social decision to be made takes into account of all individual choices.

These choice sets are called *voting-papers* on which each agent can quote his '*price*', i.e. the highest priority for additional units to a given bundle of public goods: see table 1.4.1 for an example. These prices are averaged over all agents and the average is scaled to the income in terms of money of the collectivity considered. The resulting price is called the *social benefit price* of the bundle of public goods proposed. If the social benefit price equals the social cost price, then the bundle is said to be an equilibrium bundle (see section 5.2).

Table 1.4.1. A voting-paper for a city-economy

*Public commodity**	*Proposed quantity in crucial characteristics*	*Agent's priority to one additional unit*
housing	1000 per year	...
city reconstruction	10 blocks	...
private transport + roads	1000 acres	...
public transport	every 15 minutes	...
safety	300 police officers	...
parcs and recreation	1000 acres	...
health and sport	4 swimming pools	...
community development	8 district houses	...
education	30 children per teacher	+ ...
*other: depending on task		max. 100

If the social benefit price does not equal the social cost price of that bundle, then a new bundle is proposed, until an equilibrium bundle is found. This mechanism, which is assumed to work completely analogous to the market mechanism for private goods, is called a *referendum mechanism* for public goods.

The referendum mechanism is a formal concept, just as the economic market mechanism, and will never be perfectly realized, nor perfectly applicable. But it demands from the individual agents as much information for an allocation of public goods, as the market mechanism for an allocation of private goods. However, people must have an attitude towards this mechanism, different from the competitive behavior required for a market mechanism. People must be willing to be informed and to vote, just as in political elections. The equilibrium thus obtained is called a *public equilibrium*. In real life, also, the referendum mechanism is replaced by a political mechanism which appoints people to make the decisions about public goods. However, if the political mechanism appoints people through elections, this real life solution can be considered as a two-step referendum mechanism. If a political mechanism can guarantee that each representative represents the 'average' preferences of the corresponding proportion of voters and if the referendum-mechanism is applied in the political body of representatives, then the resulting decision also meets the optimality criteria developed in welfare theory. This idea is compatible with the views of Downs, mentioned in section 1.3, and his theory can be generalized in this way.

The organization of decisions in an economy proposed in section 6.5, has some characteristics similar to the proposals of Musgrave (1959), Tinbergen (1961) and Kolm (1967). Musgrave distinguishes three government branches: the allocation branch, the distribution branch and the stabilization branch. Tinbergen distinguishes various levels of local public goods which are hierarchically ordened. This ordering is similarly determined here according to the extent (the 'locality') of public goods, and a converse ordering is given by the markets of private goods.

The referendum mechanism proposed in this study can be considered to belong to the class of voting procedures. This is one method from the several methods for allocating resources mentioned by Shubik (1970). Since the analysis of economic laws and their conditioning belongs typically to the field of political economy (cf. Lange, 1959, ch. 6), it follows that the renewed interest in allocation mechanisms can be considered as a return to the classical theory of political economy (see Morgenstern, 1972, p. 1175).

It should be stressed, finally, that every allocation mechanism must be sustained by assumptions about the agents' behavior. An agent in an economy is said to have an *individual rationality* if he chooses a maximal element in his own choice set. When agents are allowed to form coalitions in an economy, they can behave rationally in a different way. An agent is said to have a *group rationality* if he joins a coalition of agents that can offer the best opportunities to all members of the coalition. Group rationality is closely related to the concept of the core, defined in section 2.2. An allocation mechanism can have an outcome which is sustained by both types of rationality (cf. the competitive equilibrium in section 2.2), but this is not necessary (see section 5.3).

2. Equilibrium in a system of economic relations

2.1. GENERAL EQUILIBRIUM THEORY

The concept of equilibrium has received attention in economic theory since its emergence as a science, although the interest it receives is not always in a state of equilibrium (cf. Kornai's Anti-Equilibrium, 1971). Partial equilibrium theory contains studies of consumer's or producer's behavior, their strivings and constraints. In general equilibrium theory, these often conflicting behavioral patterns were brought together in one model in order to analyze compatibility of the assumptions and definitions, and to develop criteria to judge possible allocations in an economy.

Analysis of equilibrium is not only interesting in order to derive properties of equilibrium situations, it is also necessary for understanding the forces and movements causing disequilibrium. Disequilibrium and change are closely related. Change is evident in economic life, but less evident the answers to questions such as: why, how and in which direction are we changing or should we change. Before trying to answer these questions about disequilibrium, a model will be developed in which an equilibrium exists, and its properties analyzed.

What is equilibrium in a social system, such as an economic system? A *system* is said to be a set of elements and a set of (binary) relations between these elements. These elements are characterized by specific features which can leave room for variations. If all variations in elements are explained by elements and relations of the system, the system is called *closed.* The elements in economic systems have specific capacities, e.g., a particular consumptive need for a consumer, or alternative uses for a unit of capital. The relations between different elements are constraints on the elements.

For every relation in a system, a state of equilibrium can be defined. If all relations are in equilibrium, then the system is said to be in *a state of equilibrium*. The system is *in motion* if one or more of its relations is not in a state of equilibrium. If all elements of the system are changing through the motion of the system, then the motion is irreversible or non-stationary and the system is *changing* (see section 7.2).

Change of social systems is virtually accepted nowadays, but the philosophy of the 'harmonia praestabilita', the belief in the stable and natural order given to everything on earth, has left an imprint on many economic theories (see Dalmulder, 1960). If this philosophy is accepted, the models can only be in a stationary motion towards a state of equilibrium. The economic systems described in this study are not changing. This is not because of the above philosophy, but because they are complex enough and serve their purpose well without proper change and also because they are closed systems, for which change is extremely difficult to implement in formal (and not intuitive) relations. A (stationary) motion towards a state of equilibrium is probably the most one can deduce. For the time being this is sufficient.

Thus, a *general equilibrium theory* describes an economic system by means of a set of relations between entities in the system, which determine values of variables that sustain a state of equilibrium in all relations. The notion 'that a social system moved by independent actions in pursuit of different values is consistent with a final coherent state of balance, and one in which the outcomes may be quite different from those intended by the agents, is surely the most important intellectual contribution that economic thought has made to the general understanding of social processes' (Arrow and Hahn, 1971, p. 1).

The first contribution to general equilibrium theory was made by Adam Smith (1776). He recognized that 'the division of labour' and other factors of production is based on the self interest of laborers and other agents to move from low to high renumeration and thus equalizing the rate of return in a society. Besides, in the pursuit of his own interest, every individual is 'led by an invisible hand to promote an end which was no part of his intention': the public interest (vol. I, p. 400). Although Smith and other classical authors have made an important contribution to the analysis of allocation of resources, they did not recognize the fact that the demand for commodities and therefore also the equality between demand and supply is influenced by prices.

This fact was considered by later authors e.g. Cournot (1838), and integrated by Walras (1874) into an economic model. Walras succeeded in describing the first closed general equilibrium model for a competitive, private ownership economy. An equilibrium price is that vector of prices at which both consumers and producers are in a state of equilibrium and at which their total supply equals total demand. Since the mathematical tools at that time were deficient, he was not able to prove existence or stability of an equilibrium.

Although Walras defined a stationary motion for his system, he also attempted to introduce proper change in his system. This is obtained if the individual agents are able to overcome the constraints put upon them by the given resources and capacities, by changing their capacities. A model of endogenously determined economic change and growth would have emerged, distinguishable from the dynamic models in which growth is determined exogenously.

Change is firstly introduced into general equilibrium theory by Marx (1871). He drew attention to the relation between the productive capacities or forces of individuals and the organization or institutional structure in an economy. This relation is in a state of disequilibrium and generates changes in elements within the system. Through a dialectic process, the economy changes irreversibly from one stage to another.

2.2. OPTIMAL ALLOCATIONS

Economic theory is not endowed with strong criteria for optimality. The first to be mentioned is rather a necessary condition: efficiency in production. This principle was applied by Pareto (1909) to consumers in an economy. Recently, Shubik (1959) and Scarf (1962) have introduced a stronger concept from game theory, called the core of an economy, which has been developed by Edgeworth (1881).

In order to define these concepts, a description of an exchange economy E_0 is given here.

The economy consists of h *consumers*, f *producers*, and m *commodities* or economic goods, where h, f and m are finite natural numbers.

For each economic good, a unit of measurement is given to indicate a quantity of that good. Quantities of all economic goods are represented in the real euclidian m-space R^m, called the *commodity space* of the economy.

Consumers are indicated by the index $i \in H := \{1, \ldots, h\}$. Each consumer is characterized by his *consumption set* X_i, which is a relation in the commodity space specifying the inputs (positive numbers) and the outputs (negative numbers) to which the i-th consumer is constrained by his needs, capacities or otherwise, and by his *preference relation* $\precsim_i$ in his consumption set. The consumers are also endowed with a vector of original resources, $w_i \in R^m$.

The producers are indicated by the index $j \in F := \{1, \ldots, f\}$. Each producer is characterized by his *production set* Y_j, which is a relation in the commodity space specifying the outputs (positive numbers) which the producer can produce by means of certain quantities of inputs (negative numbers). The surplus of production of producer j is distributed to the consumers in the proportion θ_{ij}, with $\sum_i \theta_{ij} = 1$.

The economy E_0 is thus defined as follows:

$$E_0 := \{H, (X_i, \precsim_i, w_i);\ F, (Y_j, \theta_{ij})\}.$$

Consumers and producers are *agents* in the economy. These agents can be individual persons, but also collectivities such as households or firms. Agents make a choice from their choice sets, decide to act, and consume or produce. No distinction is made between choice, decision and activity. Many agents are both consumer and producer: in that case they have a consumer index and a producer index. Finally, an employee can be considered as a producer, or as a holder of resources, viz. labor, selling these services to a producer.

Agents choose elements from their choice sets. The choice sets of producer j is equal to his production set Y_j. The choice set of consumer i is determined by his consumption set X_i, the value of his resources w_i and his share in the profit of producers:

$$\begin{aligned} M_i(p) &:= \{x_i \in X_i \mid px_i \leqq pw_i + \sum_j \theta_{ij} py_j,\ y_j \in Y_j\} \\ &= \{x_i \in X_i \mid px_i \leqq \lambda_i(p)\}. \end{aligned}$$

This choice set $M_i(p)$ is also called the consumer's *budget set*, and is determined by the prices and the consumer's *income* $\lambda_i(p)$. If the consumers are in possession of the primary resources w_i and of the production units, then their income depends on the value of these resources and production activities, and thus on the prices. The economy E_0 in which the income distribution is determined by the initial distribution of resources and of shares in firms, and by the prices formed on a market of resources and other commodities, is called an *exchange economy*. The commodities in an exchange economy are all privately owned.

It is possible, however, to separate the income distribution from the market and to determine exogenously each consumer's income λ_i. The set of positive real numbers $\{\lambda_i | i \in H, \lambda_i > 0 \text{ and } \sum \lambda_i = 1\}$ is called an *income distribution* and denoted by $\{\lambda_i\}$. In that case, the consumers receive an income which is independent of the actual market conditions, and specification of individual resources and shares is superfluous. The economy with only private goods

$$E_1 := \{H, (X_i, \precsim_i, \lambda_i); F, (Y_j); w\},$$

in which the income distribution is given exogenously, is called a *distribution economy* (see Malinvaud, 1969b, section 5.2). In a distribution economy, markets can exist for all commodities, so that prices are formed via competitive behavior of agents, at which the quantities produced equal the quantities consumed.

In an exchange economy as well as in a distribution economy, all agents are assumed to choose maximal elements in their choice sets. The budget sets of consumers are ordered according to the preference relation $\precsim_i$, and the production sets of producers are ordered according to the profit obtained.

A *state* or an *allocation* of the economy E_0 is an $(h+f)$-tuple of points $\{(x_i), (y_j)\}$ in R^m, specifying the activity of each agent. A state $\{(x_i), (y_j)\}$ is called *attainable* if the following conditions are satisfied:
(a) $x_i \in X_i$, for every $i \in H$,
(b) $y_j \in Y_j$, for every $j \in F$,
(c) $\sum x_i \leqq \sum y_j + \sum w_i$.
Let the set of attainable states of E_0 be indicated by $A(E_0)$. On $A(E_0) \subseteq R^{(h+f)m}$ is defined the partial ordering relation $\precsim$:

$$\{(x_i^1), (y_j^1)\} \precsim \{(x_i^2), (y_j^2)\}, \text{ if } x_i^1 \precsim_i x_i^2, \text{ for all } i \in H.$$

A state is called *Pareto superior to* or *dominating* another state if it is greater according to the relation $\precsim$. A state is called *Pareto optimal* or a Pareto optimum if it is a maximal element of $A(E_0)$ for $\precsim$. The set of Pareto optima in E_0 is indicated by $B(E_0)$.

A more general concept, i.e. based on group rationality, is the core of an economy. Let $\bar{E}_0$ be an economy E_0 without production, i.e. an exchange economy. A *coalition* S is a set of consumers, $S \subseteq H$. A state (x_i) is said to be attainable for coalition S if $x_i \in X_i$, for all $i \in S$, and $\sum_{i \in S} x_i \leqq \sum_{i \in S} w_i$. A state (x_i^1) is *blocked* by coalition S if there is another attainable state (x_i^2)

for coalition S so that $x_i^1 \prec_i x_i^2$, for all $i \in S$. The *core* of economy $\bar{E}_0$ is the set of all attainable states that are not blocked by any coalition, and is indicated by $C(\bar{E}_0)$. It is evident that $A(\bar{E}_0) \supseteq B(\bar{E}_0) \supseteq C(\bar{E}_0)$.

If the production sets and preference orderings are convex, then a necessary and sufficient condition for Pareto optimality is the existence of a hyperplane separating the sum of the production sets and the sum of upper-preference sets of an allocation. This approach (described in chapter 5) is also valid if public goods are present in the economy (see Fabre, 1969).

If the production and utility functions are also differentiable, then necessary and sufficient conditions can be expressed in terms of marginal utilities and marginal productivities (see Samuelson, 1954). This approach is taken in classical welfare theory. One of the main results of welfare theory is that the existence of markets and convexity of production and preferences means that with each Pareto optimal allocation, a price can be associated that sustains the allocation. This price and the allocation is called a competitive equilibrium. (See also Arrow, 1970).

2.3. AN EXCHANGE ECONOMY

The allocation process through a price forming market mechanism with competitive behavior of all agents, is formally described by the Competitive Equilibrium models. If all resources and production units in an economy are privately owned by agents, and exchange of these resources and products determines the incomes of the agents, then the economy is called an *exchange economy with private ownership*. It has been mentioned already that Walras (1874) first succeeded in designing a competitive equilibrium for such a private ownership economy. This economy also included production, investment, and storage of products and production factors.

Properties of the market mechanism are given in sections 1.4 and 6.1. The income distribution in a private ownership economy is partially determined endogenously, because it depends not only on the given distribution of ownership, but also on the prices of these possessions. The Walrasian model has been generalized and improved greatly by later scholars, although the essential characteristics have remained unaltered.

The uniqueness of a competitive equilibrium was denied by J.G. Koopmans (1932). The existence of a competitive equilibrium was demonstrated firstly by A. Wald (1936). In the meantime, new tools were applied in economic theory (cf. T. Koopmans, 1951). An important instrument in this context was Brouwer's fixed-point theorem, applied by von Neuman (1937) to prove the existence of a process of proportional growth in a competitive economy. The set theoretical approach, the developments in linear programming, and topological results as fixed-point theorems, renewed interest both in optimality problems and in general equilibrium theory. These new formulations also brought out clearly 'the basic unity of welfare economics with the descriptive theory of competitive equilibrium' (Koopmans, 1957, p. 6).

With these new tools, a competitive equilibrium for a more general economy can be shown to exist. This was done firstly by McKenzie (1954) and Arrow and Debreu (1954). A complete and translucent exposition of a private ownership economy, rigorously built on an axiomatic foundation, is given by Debreu (1959). An outline of this model will be given below. A more general model is given by Debreu (1962).

Competitive equilibria are based upon price adapting behavior of agents. Another solution for the allocation of resources in an exchange economy is based upon coalition forming behavior of agents. In that case, it is assumed that every agent can freely form coalitions with other agents and will do that if it is advantageous for every member of the coalition. The set of coalition equilibria is called the core of an economy (see the previous section). It was shown by Debreu and Scarf (1963) that a competitive equilibrium determines an allocation which also belongs to the core of the economy, and that the core converges to the set of competitive equilibria if the number of agents becomes very large. This was conjectured already by Edgeworth (1881) for a two person economy, and is caused by the fact that the influence of an individual agent on the final allocation is negligible. A precise mathematical formulation of such a situation is given by Aumann (1964), who defined the set of economic agents to be an atomless measure space, called a *continuum of agents*, and by Vind (1964). Thus, for an exchange economy with a continuum of traders it is possible to find a price vector that sustains an allocation in the core, without changing the initial allocation of resources. In that case it is also possible to relax some of the conditions for the existence of an equilibrium (see Hildenbrand, 1970). The economics described in this study, however, are assumed to consist of a finite number of agents.

Consider the exchange economy for private goods with private ownership $E_0 := \{H, (X_i, \precsim_i, w_i); F, (Y_j, \theta_{ij})\}$. An allocation $\{(\tilde{x}_i), (\tilde{y}_j)\}$ and a price vector $\tilde{p}$ is said to be a *competitive equilibrium* if:

(a) Every consumer $i \in H$ is in a state of equilibrium, i.e. $\tilde{x}_i$ is a maximal element in $\{x_i \in X_i \mid \tilde{p}x_i \leqq \tilde{p}w_i + \sum_j \theta_{ij}\tilde{p}\tilde{y}_j\}$ for $\precsim_i$.

(b) Every producer $j \in F$ is in a state of equilibrium, i.e. $\tilde{p}\tilde{y}_j$ is a maximal element in $\{\tilde{p}y_j | y_j \in Y_j\}$.

(c) The market is in equilibrium, i.e.
$$\sum_i \tilde{x}_i = \sum_j \tilde{y}_i + \sum_i w_i.$$

Let the economy E_0 satisfy the following conditions:

Assumption 2.3.1. $0 \in Y_j$, i.e.
each firm has the possibility of inaction if loss would otherwise be incurred.

Assumption 2.3.2. $\sum Y_j$ is closed, i.e.
the social production set $Y := \sum Y_j$ contains the production vector which is the limit of a sequence of production vectors in Y.

Assumption 2.3.3. $Y \cap (-Y) \subseteqq \{0\}$, i.e.
the social production is irreversible, or if $y \neq 0$ and $y \in Y$, then $-y \notin Y$.

Assumption 2.3.4. $R^n_- \subseteq Y$, i.e.
any commodity can freely be disposed of.

Assumption 2.3.5. Y *is* convex, i.e.
every production vector can be produced which belongs to the line segment connecting two producible production vectors.

These assumptions about production are sometimes rough approximations of reality, e.g. if some commodities are socially produced with increasing returns to scale or with indivisibilities. The next five assumptions are about the consumption characteristics for each consumer $i \in H$.

Assumption 2.3.6. X_i is closed, i.e.
each consumer's consumption set contains the consumption vector which is the limit of a sequence of consumption bundles in X_i.

Assumption 2.3.7. X_i is bounded below for $\leqq$, i.e.
each consumer can assign a lower bound for every commodity, whether it is positively or negatively appreciated.

Assumption 2.3.8. X_i is convex, i.e.
each consumer can appreciate a bundle which belongs to the line segment connecting two bundles in his consumption set.

Assumption 2.3.9. There exists an $x_i \in X_i$ such that $x_i < w_i$, i.e. each consumer can dispose of initial resources above his minimum level of subsistance.

The last assumption is made about the preference relation $\lesssim_i$ on X_i. Therefore, with each $x_i \in X_i$ is associated the set of consumption bundles $C_i(x_i) := \{y_i \in X_i \mid x_i \lesssim_i y_i\}$, called an *upper-preference set.*

Assumption 2.3.10. The preference relation $\lesssim_i$ on X_i is:
(a) transive and complete (see section 8.1),
(b) continuous : $C_i(x_i)$ and $C_i^{-1}(x_i)$ are closed in X_i,
(c) convex : $C_i(x_i)$ is convex,
(d) without satiation: there exists no greatest element in X_i for $\lesssim_i$.

This assumption demands a rational and diligent consumer, who can overlook and order all his consumption possibilities for every imaginable quantity, and who is not influenced by the consumption of other people. These rigorous requirements are needed to determine an activity in all, even extreme and improbable, situations which may arise in the model.

In order to appreciate the concepts and proofs for an economy with public goods, some concepts for a private ownership economy are given here, as well as an *outline* of the proof of existence of a competitive equilibrium.

An important problem for the proof is the unboundedness of both production and consumption sets. This can cause a lack of continuity in the demand or the supply of consumers and producers as a function of the prices. At one price the demand can be infinite and at another – arbitrarily close – price, the demand can be finite. In the course of the proof however, it is shown (see Debreu, 1959, p. 85) that the relevant subsets of the production and consumption sets are bounded. In order to simplify this reasoning, it is assumed here that the sets X_i and Y_j are bounded to their relevant subsets $\overline{X}_i$ and $\overline{Y}_j$ (and assumptions 2.3.4 and 2.3.10d are temporarily neglected). Then the following concepts and properties can be defined and derived.

For each producer $j \in F$, a *profit function* $\pi_j : R^m \to R$ is defined by $\pi_j(p) := \max\ py$, with $y \in Y_j$. The producer will choose that activity which maximizes his profit. Since this result can be obtained from more than one production activity (contrary to the classical calculus approach with strict convexity requirements), a producer's *supply multifunction* y_j :

$R^m \to \overline{Y}_j$ is defined by

$$y_j(p) := \{y \in \overline{Y}_j \mid py = \pi_j(p)\}.$$

The sum of the producer's supply multifunctions is called the *social supply multifunction* $y(p) := \sum y_j(p)$. From properties 8.2.4 and 8.2.3 it follows that this multifunction is upper hemi-continuous.

Next, the consumer's *budget multifunction* $M_i : R^m \to \overline{X}_i$ is defined by (see the previous section):

$$M_i(p) := \{x \in \overline{X}_i \mid px \leqq pw_i + \sum_j \theta_{ij} \pi_j(p)\}.$$

The assumptions imply that for each p, the set $M_i(p)$ is non-empty, convex and compact, and that M_i is continuous on R^m. Each consumer chooses a greatest element in his budget set, which ensures a state of equilibrium relative to the price p. The consumer's *demand multifunction* $x_i : R^m \to \overline{X}_i$ is thus defined by:

$$x_i(p) := \{x \in M_i(p) \mid \text{for all } y \in M_i(p) : y \lesssim_i x\}.$$

The sum of the consumer's demand multifunctions $x(p) := \sum_i x_i(p)$ is called the *social demand multifunction.*

Finally, the difference between demand and supply in a market is determined by the *excess demand multifunction* $z : R^m \to \overline{Z}$, defined by:

$$z(p) := x(p) - y(p) - \{\sum w_i\}.$$

From the above definition of a competitive equilibrium it can be deduced that the price vector $\tilde{p}$ sustains a competitive equilibrium if and only if $0 \in z(\tilde{p})$. Therefore, the problem of existence is solved if there exists a price $\tilde{p}$ such that $0 \in z(\tilde{p})$.

Since $x(p)$ and $z(p)$ are upper hemi-continuous, and non-empty, convex and compact foı each p, and since the price space can be reduced to a compact subset $\overline{P} \subseteq R^m$ without loss of generality by fixing the sum of all pıices, it is possible to define a multifunction φ from $\overline{Z} \times \overline{P}$ into itself which satisfies the conditions of Kakutani's fixed point theorem (10.2.1). A fixed point $(\tilde{z}, \tilde{p})$ is thus shown to exist and the definition of φ implies that $0 \in z(\tilde{p})$.

The concepts mentioned here will be analogously defined in section 5.2 from the quantity space into the price space, in order to determine the social benefit and social cost of a certain bundle of public goods.

It is easily checked that the allocation $\{(\tilde{x}_i), (\tilde{y}_j)\}$ is Pareto optimal. If it were not Pareto optimal, then a Pareto superior allocation $\{(x_i), (y_i)\}$ would exist such that:

for all $j \in F$: $-\tilde{p}\tilde{y}_j \leqq -\tilde{p}y_j$ with $y_j \in Y_j$,
for all $i \in H$: $\tilde{p}\tilde{x}_i \leqq \tilde{p}x_i$ with $\tilde{x}_i \lesssim_i x_i$, and
for some $h \in H$: $\tilde{p}\tilde{x}_h < \tilde{p}x_h$ with $\tilde{x}_h \prec_h x_h$.
Thus $\tilde{p}(\sum \tilde{x}_i - \sum \tilde{y}_j) < \tilde{p}(\sum x_i - \sum y_j) = \tilde{p}(\sum w_i)$.

This contradicts part (c) of the definition of a competitive equilibrium, requiring that $\sum \tilde{x}_i - \sum \tilde{y}_j = \sum w_i$.

Similar results can be obtained for a distribution economy E_1 with private goods and individual decisions about consumption and production (see property 5.3.4).

3. Production

3.1. PRODUCTION SETS AND PRODUCTION MULTIFUNCTIONS

Each producer in an economy can be definition decide about the quantities of output of the commodities he wants to produce, which are in turn related to the quantities of input of the production factors he needs for production. The set of all input-output combinations for a producer is called the producer's production set (see 2.2), and gives full information about the producer's technological production possibilities. It is, however, not the only way of describing production possibilities; this can be done equivalently by means of a relation (a function or multifunction) defined on the space of economic goods.

Assume that there are l commodities in the economy; then any specification of quantities of these commodities is an element of the l-dimensional euclidean vector space R^l. A component z_i of $z \in R^l$ is said to be an *input* if it is negative and an *output* if it is positive.

Choose any bipartition of the set of commodities, say m and n, such that $l = m + n$ and $m \neq 0$, $n \neq 0$. This bisection can be done according to any criterion, although it is useful to anticipate the problem under consideration. In most cases, one assumes that the input or output characterization is tied to a specific commodity and the bisection goes along these lines (see the examples below). In this study, the bisection is also done according to the private or public character of the economic goods.

Given the bisection of R^l into R^m and R^n, and the production set $Y \subseteq R^l$, the *input-output multifunction* $f: R^m \to R^n$ is defined by $f(x) := \{y \mid (-x, y) \in Y\}$. On the other hand, given the input-output multifunction f, the *production set* Y is defined by $Y := \{(-x, y) \mid y \in f(x)\}$.

A typical production set Y in two dimensions is given in fig. 3.1.1; the commodity x is an input-good ($x \leqq 0$) and the commodity y is an output-good (there exists a $y > 0$). This production set Y is extended to a production set Y_1 with two inputs, x_1 and x_2, in fig. 3.1.2.

Fig. 3.1.1. A production set Y.

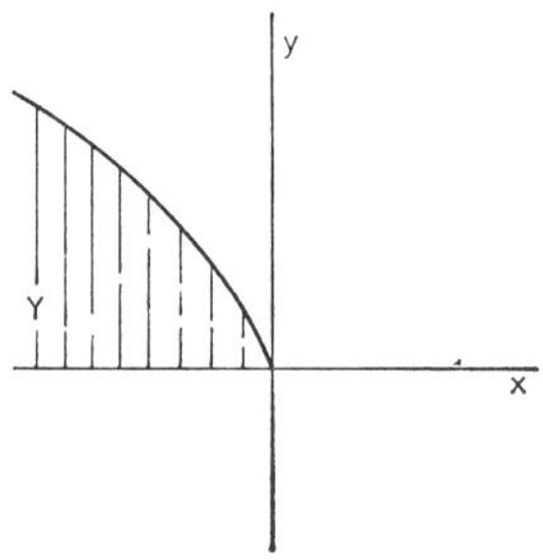

Fig. 3.1.2. A production set Y_1 with two inputs.

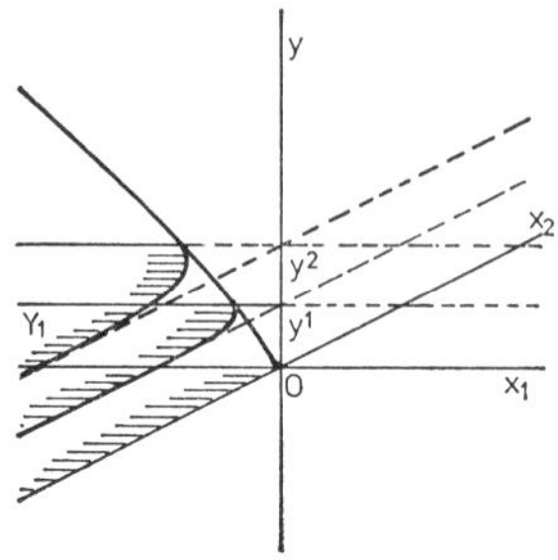

The input-output multifunction is defined such that a component x_i of $x \in D(f)$, i.e. the domain of f, is an input if it is positive and an output if it is negative, and such that a component y_i of $y \in R(f)$, i.e. the range of f, is an output if it is positive and an input if it is negative.

If, however, the technology is such that the commodity space R^l can be bisected into a factor-, or input-, space R^m and a product-, or output-, space R^n, and if the production multifunction f is defined from R^m into R^n, then all $x \in D(f)$ and $y \in R(f)$ are nonnegative. Production multifunctions are usually assumed to be nonnegative, implying the bisection of commodities in inputs (or factors) and outputs (or products). Unless otherwise stated, this assumption will also be made here.

The set-representation of the production technology Y_1 in fig. 3.1.2 may be replaced by the representation of the input sets belonging to a certain level of output. These sets are called *input sets* (or level sets) and specify the combination of factor-input quantities which enable the producer to produce at least a given quantity of output. The boundaries of these factor sets are called the *isoquants* corresponding to a given output- or production-level, and are denoted by $x(y) := \mathrm{Bd} f^{-1}(y)$ (see fig. 3.1.3).

The boundaries of the factor sets (and of all other sets in this study) are of crucial importance. It will be seen that the valuation of elements in the set is determined by the elements on the boundary of the set. The economic interpretation of this fact is that it is only these latter elements which are optimally used, or are in some sense 'efficient'. An element of a set is said to be *efficient* if it is a maximal or minimal element of that set, according to a given ordering relation. Let the ordering relation be

Fig. 3.1.3. Input-isoquants related to Y_1 in fig. 3.1.2.

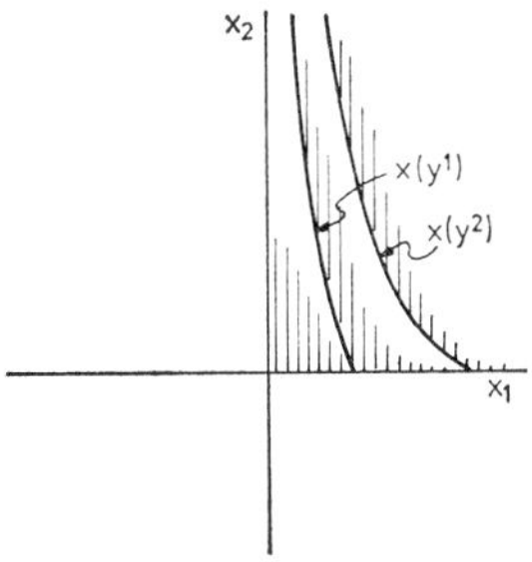

'less is better', or: $x^1 \leqq x^2$ is equivalent to $x^1 \gtrsim x^2$. Then an efficient element of a factor set is a maximal element in the sense that there does not exist an element that produces the given output-level with less factor-input in any component. Another ordering relation is 'closer to zero is better', or: $x^1 = \lambda x^2$ and $0 \leqq \lambda \leqq 1$ is equivalent to $x^1 \gtrsim x^2$. Then no proportional diminution of an efficient factor-input is possible if one wants to maintain the required output level.

In both cases, the set of efficient points of a factor set, $f^{-1}(y)$, is contained in the boundary of that set, $x(y)$. It should be noticed that the set of efficient points according to the criteria mentioned above overlap if the factor sets are both monotonous and aureoled; this is the case in fig. 3.1.3.

Another extension of the production set Y in fig. 3.1.1 is obtained by adding an output to the production set, to get the production set Y_2 with outputs y_1 and y_2 and with input x (see fig. 3.1.4). The production technology can also be represented by the corresponding input-output multifunction $f: R \to R^2$ defined by $f(x) := \{(y_1, y_2) \mid (-x, y_1, y_2) \in Y_2\}$. This is done by means of the boundaries of the *output-sets* $f(x)$, which boundaries are called the *product transformation curves* or *production-possibility frontiers* corresponding to a given quantity of factor-inputs (see fig. 3.1.5), and which are denoted by $y(x) := \mathrm{Bd}\, f(x)$.

Again it is true that all efficient points of the output-sets $f(x)$ are elements of the boundary of the output-sets. These efficient points are minimal elements according to the ordering relations defined above, or maximal elements according to the ordering relations 'more is better', i.e. $x^1 \leqq x^2 \Leftrightarrow x^2 \lesssim x^2$, and 'farther from zero is better', i.e. $x^1 = \lambda x^2$ and $\lambda \geqq 1 \Leftrightarrow x^1 \gtrsim x^2$. If the product sets are both lower-monotonous and starred (as in fig. 3.1.5) then the sets of efficient points coincide.

Fig. 3.1.4. A production set Y_2 with two outputs.

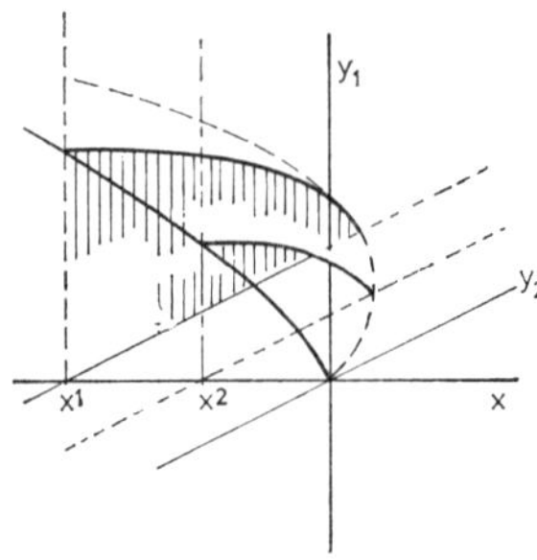

Fig. 3.1.5. Output-isoquants related to fig. 3.1.4.

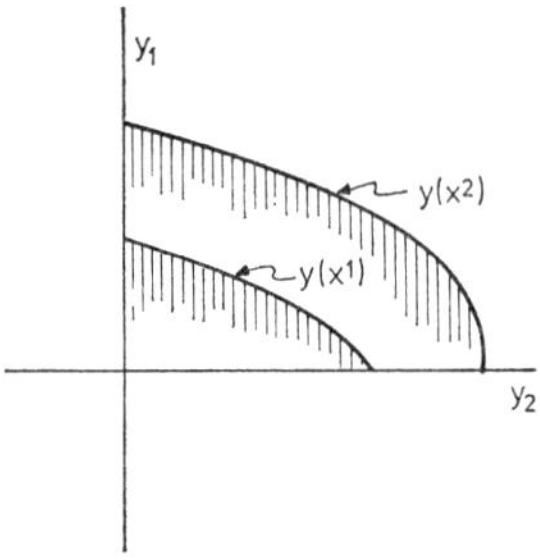

The set of efficient points contains only one element if the output of the input-output multifunction consists of only one commodity. The function which assigns to every input-vector the (unique) maximal or efficient element of the corresponding one-commodity output-set is said to be a *production function*. Production functions are frequently used to describe a production technology, but not in this study. The reason for this is, of course, that they can only be used when the technology under consideration has a one-commodity output. The conversion of an input-output multifunction $f: R^m \to R_+$ into a production function $\bar{f}: R^m \to R_+$ and vice versa is accomplished by:

$$\bar{f}(x) := \max\{f(x)\}, \text{ resp. } f(x) := \{y \mid y \leqq \bar{f}(x)\}.$$

Finally, the inverse of an input-output multifunction $f: R^m \to R^n$ is called the *output-input multifunction* $f^{-1}: R^n \to R^m$. It follows then that:

$$f^{-1}(y) = \{x \mid y \in f(x)\} = \{x \mid (-x, y) \in Y\}.$$

3.2. CONDITIONS IN THE PRODUCTION MODEL

Axioms about the production technology can be formulated in terms of conditions on the production set Y in the commodity space (defined in 2.2), or on the input-output multifunction (defined in 3.1). Because the one can be derived from the other; the conditions are narrowly related. Still, conditions on multifunctions give more opportunity for gradations than do conditions on sets.

Using results obtained in chapter 10, the above statement can be made more precise. Firstly, the following axioms or assumptions about production *sets* will be defined (see also section 2.3):

Let Y be a production set in the commodity space $R^l = R^m \times R^n$.
A1. Y is closed;
A2. $Y \cap -Y = \{0\}$; (irreversible production)
A3. Y is unbounded;
A4. $Y \supset R^l_-$; (free disposal)
A5. $\text{Proj}_1\ Y = R^m_-$ and $\text{Proj}_2\ Y \subseteq R^n_+$; (input-output partition)
A6. Y is convex.

Assumptions A1–A6, except for A4, are met by the production sets in figs. 3.1.1, 3.1.2 and 3.1.4. If A4 is chosen, A3 may be dropped and A5 must be dropped. Let f: $R^m \to R^n$ be an input-output multifunction. Then the following assumptions may be defined (see also sections 8.2 and 10.2):
B1. f has a closed graph, $G(f)$;
B2. $f(x)$ is upper bounded for every $x \in D(f)$;
B3. $R(f) = D(f^{-1})$ is a cone and there exists a non-zero $x \in D(f)$ such that $f(0) \subset f(x)$:
B4. f is starred, i.e. f is point-starred and f^{-1} is point-aureoled;
B5. f is monotone increasing, i.e. f^{-1} is monotone decreasing;
B6. f is non-negative;
B7. f and f^{-1} are quasi-convex;
B8. f is convex;
B9. f is a convex star process;
B10. f is homogeneous of degree k.

Shephard (1970, p. 185) defines an input-output multifunction (or production correspondence) in terms of the assumptions B(1, 2, 3, 5, 6, 7). In a 1972 paper, however, he uses only the assumptions B(1, 2, 3, 4, 6) to define an input-output multifunction.

It can be easily checked that both series of assumptions imply assumptions A(1, 2, 3, 5) about production sets, but that the converse is not true. Further, B(6) is equivalent to A(5) and B(8) is equivalent to A(6).

If one assumes B(9), then all other assumptions are also implied except for B(5 and 6). Assumption B(9) also implies A(1, 2, 3, 6).

Property 3.2.1. (equivalences between Y and f)

Let $Y \subseteq R^{m+n}$ be a production set and f: $R^m \to R^n$ be the corresponding input-output multifunction. Then the following pairs of statements are equivalent:

1.a. Y is closed and convex, i.e. A(1, 6);
   b. f is a convex process, i.e. B(1, 8).
2.a. Y is closed, convex and contains R^n_-, i.e. A(1, 4, 6);
   b. f is a monotone increasing convex process, i.e. B(1, 5, 8).
3.a. Y is closed, convex and input-output partitioned, i.e. A(1, 5, 6);
   b. f is a non-negative starred convex process, i.e. B(1, 4, 6, 8).

Proof

1. As $Y = \{(-x, y) \mid y \in f(x)\}$, the equivalence follows by definition.
2. As Y is closed and convex, f is a convex process.
   Further, as Y is starred ($0 \in Y$ and Y convex), it follows from property 8.3.1 that the recession cone $0^+ Y = \text{Conint } Y \supseteq R^n_-$. Therefore, Less $Y = \{z \mid \exists \bar{z} \in Y : z \leqq \bar{z}\} = Y$, which is equivalent to f being monotone increasing, according to property 10.2.5.1.
3. As Y is closed and convex, f is a convex process.
   As Y is input-output partitioned, i.e. $(-x, y) \in Y$ implies that $-x \leqq 0$ and $y \geqq 0$, f is non-negative.
   As $\text{Proj}_1 \; Y = R^l_-$, it follows that $(-x, 0) \in Y$, for each $x \in D(f)$; as $0 \in f(x)$ and $f(x)$ is convex, f is point-starred. Further, assume that $x \in f^{-1}(y)$ and $\lambda x \notin f^{-1}(y)$, for some $\lambda > 1$. Then $y \in f(x)$ and $y \notin f(\lambda x)$. This contradicts the fact that $0^+ Y \supseteq \{(-x, 0) \mid -x \in R^l_-\}$, such that $(-x, y) \in Y$ and $(\lambda - 1) \geqq 0$ imply $(-x, y) + (\lambda - 1)(-x, 0) = (-\lambda x, y) \in Y$. Therefore f^{-1} is point-aureoled and, by definition, f is starred.
4. As f is a convex process, Y is closed and convex.
   As f^{-1} is point-aureoled, Y is unbounded.
   As f is point-starred and $0 \in f_c(0) = f(0)$, $G(f)$ is contained in a pointed cone and $G(f) \cap -G(f) = \{0\}$. □

Some features of a technology however cannot be so easily expressed in terms of sets: e.g.,
f being non-negative monotone increasing;
f being starred;
f being quasi-convex;
f being homothetic or homogeneous, etc.

It should be noticed that assumptions A(1, 2, 4, 6) about the production set Y are frequently used (compare Debreu, 1959, p. 85, and section 2.3 above). These assumptions imply that the corresponding input-output multifunction $f: R^m \to R^n$ is a monotone increasing convex process such that $f(0)$ and $f^{-1}(0)$ are cones. The properties of convex processes (see section 10.3) are thus also applicable.

The assumptions about the input-output multifunction have, of course, implications for the inverse, the output-input multifunction. This is formulated in the following theorem, which follows immediately from the definitions previously given (see sections 10.2 and 10.5):

Property 3.2.2. (*equivalent assumptions on the inverse*)

Let $f: R^l \to R^m$ be an input-output multifunction, and $f^{-1}: R^m \to R^l$ its inverse, the output-input multifunction. Then:

1. $f^{-1}(y)$ is lower-bounded for every y, if and only if $f(x)$ is upper-bounded for every x (B2);
2. f^{-1} is aureoled, if and only if f is starred (B4);
3. f^{-1} is monotone decreasing, if and only if f is monotone increasing (B5);
4. f^{-1} is a convex aureole process, if and only if f is a convex star process (B9);
5. f^{-1} is homogeneous of degree $1/k$, if and only if f is homogeneous of degree k (B10).

The other assumptions are equally valid.

3.3. THE PRICE STRUCTURE CORRESPONDING TO A TECHNOLOGY

The technology has implications for the various representations given in terms of quantities, such as the production set, the input-output multifunction and the output-input multifunction. Under the assumption that prices are determined by marginal conditions, it will be shown that technology also has implications on the prices feasible for a given technology. Moreover, technology can be represented in terms of prices, exactly as it was shown that it was possible to represent technology in terms of quantities.

Two kinds of effects of technology on prices are studied here. Firstly, the effect on prices of outputs, (resp. inputs), if the *turnover* is given; it will be shown that this effect can be represented by the polar multifunctions of f (resp. f^{-1}) defined from R^m into R^{n*} (resp. from R^n into R^{m*}).

Secondly, the effect on the prices of outputs and inputs, if the *profit* is given; this effect can be represented by the adjoint multifunctions of f (resp. f^{-1}) defined from R^{n*} into R^{m*} (resp. from R^{m*} into R^{n*}). The definitions of polar- or adjoint multifunctions are given in section 10.1. Before the relevant properties are deduced, the following example is given to illustrate the above points.

Consider the function $\bar{f}: R^2_+ \to R_+$, defined by $\bar{f}(x_1, x_2) := x_1^\alpha x_2^\beta$. This function belongs to the well-known class of Cobb-Douglas functions, when α and β are positive. From this function $\bar{f}$, a multifunction $f: R^2_+ \to R_+$ can be derived by positing that if $\bar{y} = \bar{f}(\bar{x}_1, \bar{x}_2)$, then $\bar{y} \in f(\bar{x}_1, \bar{x}_2)$ and all y such that $0 \leqq y \leqq \bar{y}$. The multifunction f is called the non-negative less-closure of $\bar{f}$ (see sections 8.2), and belongs to the class of Cobb-Douglas multifunctions (see fig. 3.3.1).

The following three assumptions are made:

Fig. 3.3.1. The graph of a Cobb-Douglas input-output multifunction and its adjoint.

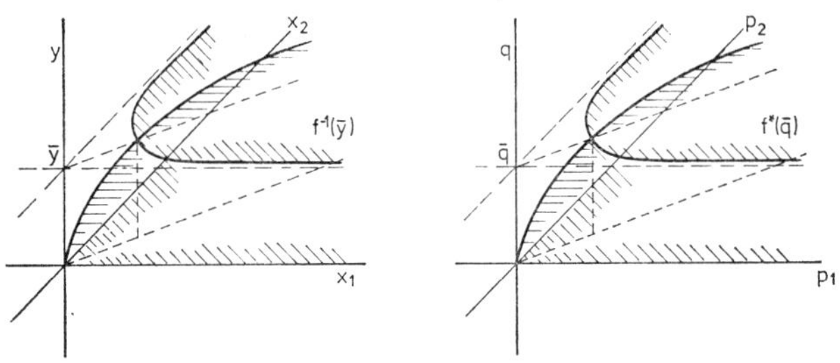

Fig. 3.3.2. An input set corresponding to output $\bar{y}$ and its polar set, the input-cost set.

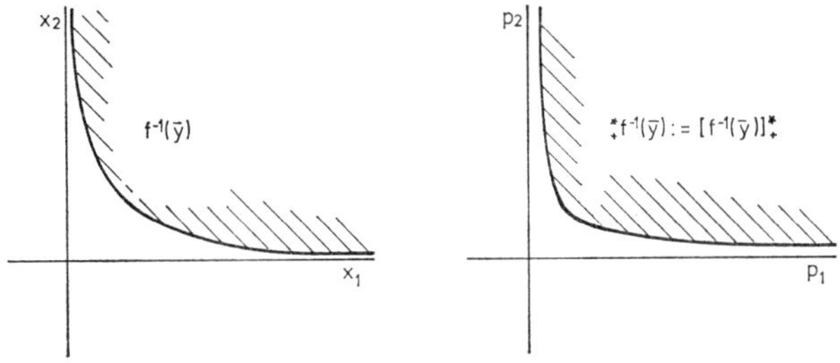

1. Let the input-output multifunction $f\colon R^2_+ \to R_+$ be defined by the Cobb-Douglas multifunction $f(x_1, x_2) := \{y \mid y \leqq x_1^\alpha x_2^\beta\}$.
We consider first the case of decreasing returns of scale, i.e. $\alpha+\beta<1$. The graph of this multifunction, $G(f)$, is depicted in fig. 3.3.1.

2. Let the prices be proportional to the marginal productivities of the inputs, resp. to the marginal costs of the outputs. This can be realized under a régime of perfect competition.
Then the solution of the problem:
maximize the value $(qy - p_1 x_1 - p_2 x_2)$
under the constraint $y = x_1^\alpha x_2^\beta$
is: $x_1 = \alpha qy/p_1$ and $x_2 = \beta qy/p_2$.
It also follows that:

$$p_1 x_1 + p_2 x_2 = (\alpha+\beta) qy.$$

3a. Next, let the value of the input for a production activity be given and scaled to the value 1, i.e. total expenditures $p_1 x_1 + p_2 x_2 = 1$, for $y = x_1^\alpha x_2^\beta$. It follows that $qy = (\alpha+\beta)^{-1}$, and some substitution of x_1 and x_2 results in: $y = (\alpha+\beta)^{-(\alpha+\beta)} (\alpha/p_1)^\alpha (\beta/p_2)^\beta$.
Thus, the output is expressed in terms of input prices or – conversely – with each output a set of input prices can be associated that gives the input a value of 1.

With each output $\bar{y}$, therefore, set of input prices can be associated that gives the input a value of at least 1:

$$\{(p_1, p_2) \in R^2_+ \mid (\alpha+\beta)^{-(\alpha+\beta)} (\alpha/p_1)^\alpha (\beta/p_2)^\beta \leqq \bar{y}\}.$$

This set can also be derived as follows. The set of inputs producing at least the output $\bar{y}$ is given by the inverse of f:

$$f^{-1}(\bar{y}) = \{(x_1, x_2) \mid \bar{y} \leqq x_1^\alpha x_2^\beta\}.$$

The upper-polar set of this, defined in section 10.1, is the set of input prices assigning to each input in $f^{-1}(\bar{y})$ a value of at least 1:

$$\begin{aligned}[f^{-1}(\bar{y})]^*_+ &= \{(p_1, p_2) \mid \forall (x_1, x_2) \in f^{-1}(\bar{y})\colon p_1 x_1 + p_2 x_2 \geqq 1\}\\ &= \{(p_1, p_2) \mid (\alpha+\beta)^{-(\alpha+\beta)} (\alpha/p_1)^\alpha (\beta/p_2)^\beta \leqq \bar{y}\}.\end{aligned}$$

The input set $f^{-1}(\bar{y})$, and the input-price set $[f^{-1}(\bar{y})]^*_+$ are represented in fig. 3.3.2. The multifunction ${}^*f^{-1}\colon R^n \to R^{m*}$ generating input-price sets is called an *output-input-price multifunction* and is defined by

$${}^*f^{-1}(y) := [f^{-1}(y)]^*_+, \qquad \text{for every } y \in \text{Range } f.$$

3b. Finally, let the value of the output for a production activity be given and scaled to the value 1, i.e. total income $qy=1$, for $y=x_1^{\beta}x_2^{\alpha}$. Then $q=x_1^{-\alpha}x_2^{-\beta}$, and with each input $(\bar{x}_1, \bar{x}_2)$ can be associated a set of output-prices giving the income a value of at most 1. This set is equal to the lower polar set of $f(\bar{x}_1, \bar{x}_2)$:

$$\begin{aligned}[f(\bar{x}_1, \bar{x}_2)]^*_- &= \{q \mid \forall y \in f(\bar{x}_1, \bar{x}_2) : qy \leqq 1\} \\ &= \{q \mid q \leqq x_1^{-\alpha}x_2^{-\beta}\}.\end{aligned}$$

The multifunction $^*f: R^m \to R^{n*}$ generating output-price sets is called an *input-output-price multifunction* and is defined by

$$^*f(x) := [f(x)]^*_-, \text{ for every } x \in \operatorname{Dom} f.$$

It may be noticed that for some feasible production plan $(\bar{x}_1, \bar{x}_2, \bar{y}) \in G(f)$ the value of total income at prices $\bar{q} \in ^*_- f(\bar{x}_1, \bar{x}_2)$ just equals the value of total expenditure at prices $(\bar{p}_1, \bar{p}_2) \in ^*_+ f^{-1}(\bar{y})$, also in the case of decreasing returns of scale. This is due to the fact that the value of inputs and the value of outputs are independently scaled to the value 1.

3′. In order to determine the prices of inputs and outputs simultaneously, thereby recognizing the fact that in the case of decreasing returns to scale total income is larger than total expenditure, we need to fix the profit value and scale it to the value 1. Instead of assumption (3) above, suppose that $qy - p_1x_1 - p_2x_2 = 1$.

As $p_1x_1 + p_2x_2 = (\alpha+\beta)\, qy$, it follows that $qy(1-\alpha-\beta) = 1$. Substitution of the values x_1 and x_2 above in $y = x_1^{\alpha}x_2^{\beta}$ and elimination of y gives:

$$q = (1-\alpha-\beta)^{(\alpha+\beta-1)}\,(p_1/\alpha)^{\alpha}\,(p_2/\beta)^{\beta}.$$

This means that the input-output multifunction

$$f(x_1, x_2) = \{y \in R_+ \mid y \leqq x_1^{\alpha}x_2^{\beta}\}$$

has the following adjoint:

$$\begin{aligned}f^*_+(q) &= \{(p_1, p_2) \mid \forall (x_1, x_2), \forall y \in f(x_1, x_2) : p_1x_1 + p_2x_2 \geqq qy - 1\} \\ &= \{(p_1, p_2) \mid q \leqq (1-\alpha-\beta)^{(\alpha+\beta-1)}\,(p_1/\alpha)^{\alpha}\,(p_2/\beta)^{\beta}\}.\end{aligned}$$

Both multifunctions are given in fig. 3.3.1.

The adjoint $f^*_+ : R^{n*} \to R^{m*}$ of an input-output multifunction $f: R^m \to R^n$, is called a *price-cost multifunction*; at output-prices q, the input-prices must be elements of $f^*_+(q)$ to ensure a profit equal to or smaller than 1.

The inverse, $f^{-1*}_- : R^{m*} \to R^{n*}$, is called a *cost-price multifunction.*

Next, consider the case of a Cobb-Douglas function $y := x_1^{\alpha}x_2^{\beta}$ with constant returns of scale, i.e. $\alpha+\beta=1$. In this case, the input-output price multifunction is the same as the one above, but the output-input, price multifunction reduces to:

$$*f^{-1}(y) = \{(p_1, p_2) \mid y \geqq (\alpha/p_1)^{\alpha} ((1-\alpha)/p_2)^{1-\alpha}\}.$$

As $p_1 x_1 + p_2 x_2 = qy$, no profits are made, and the price-cost multifunction reduces to:

$$f^*(q) = \{(p_1, p_2) \mid q \leqq (p_1/\alpha)^{\alpha} (p_2/(1-\alpha))^{1-\alpha}\}.$$

In this case, the graphs of f and f^* are cones.

As a final remark, it may be mentioned that the definitions of $*f$ and f^* can be applied to any input-output multifunction. It does not need to be continuous, differentiable, concave, etc., as is the Cobb-Douglas function.

The properties of the polar $*f$ and adjoint f^* are given in chapter 10.

Although profit is scaled to the value 1, this value 1 is not so important for the results obtained from the price-cost multifunction. It is the relative prices between the input and output components that counts. If this production model is part of a general model in which prices are equalized, then the degree of freedom obtained in scaling the value of profit to 1 is lost. Profit will then be equal to some non-negative scalar α.

One important conclusion which may be drawn from the price-cost multifunction is that profit is *not* a remuneration for managing the production process, but (at most) a reward for selecting (or developing) the production technology. Given an implementable technology, i.e. a production process with prices the economy is willing to accept, then profits are fixed under perfect competition. These profits are quite different for different production units, and need not be zero.

Profits may, however, be levelled (and approach zero) if one production unit can be split up in more units with the same technology, as in the case of decreasing returns to scale. The marginal value of output of a subunit is smaller than the marginal value of output of the integral unit. If sufficiently divided, the sum of the sub-units has a technology which can be represented by a cone (i.e. the cone closure of the technology of the integral unit), implying that each unit has zero (and equal) profit.

Thus, if all producers have the same (convex) technology and each producer can freely enter the market, then the joint technology reduces to a cone (and profit reduces to zero for each producer).

We will see that, ultimately, it is aggregate production and consumption which determine the relative prices of commodities in an economy.

The assumption of free entry into an economy, however, implies an aggregate technology (a cone) which is different from the aggregate technology when free entry is not possible in an economy.

Therefore, both the organizational structure and the individual technologies determine the social production technology and the relative prices in the economy. Assumptions about the organizational structure (such as perfect competition; free entry) cannot be omitted in a theory of value.

3.4. CONDITIONS IN THE PRICE SPACE

The assumptions about the production model were given in terms of quantities (section 3.2); these have of course, implications for the properties of dual models of the same technology, expressed in terms of prices and defined in section 3.3. Information about a production technology can be transmitted in terms of quantities, as usually is done. But is it also possible to transmit information about a technology in terms of the prices or valuations related to that technology?

The prices related to a production technology are determined by the two duality operations defined in section 3.3. Duality operations, such as those outlined in chapter 9, transform information about a technology from one space into another; in the case under consideration here: from the quantity space into the price space, and vice versa. The question is whether any information about the technology is lost in the process of transforming it by means of a duality operation. The answer depends, of course, on the properties (and complexity) of the technology.

If the assumptions about a production model are such that it is possible to regain the original model after application of duality operations, the model is called *reflexive*. No information about a technology will be lost under a duality operation if, and only if, the model describing the technology is reflexive.

From properties 9.2.3.5 and 10.3.1.6 we know that a production set Y containing zero is reflexive under the duality operation * if, and only if, Y is closed and convex; if Y is a convex cone, then Y is also reflexive under the polarity operation 0, defined in section 9.2. An input-output multifunction $f: R^m \to R^n$ is reflexive under the operation * if and only if it is a convex process; if f is a convex cone process (see section 10.3), then f is also reflexive under the polarity operation 0, defined in 10.1.

The next question is: what are the properties of the model in the price space given some specific assumptions about the model in the quantity space?

Let us consider firstly the effects of the technology on input prices *or* output prices, in the case of the polarity operation.

Property 3.4.1

Let $f: R^m \to R^n$ be either a starred input-output multifunction or an aureoled output-input multifunction; $*f: R^m \to R^{n*}$ is then a point-starred input-output price multifunction, resp. a point-aureoled output-input price multifunction. The following pairs of conditions are equivalent:

1.a. f is homothetic;
  b. $*f$ is homothetic.

2.a. f is homogeneous of degree k;
  b. $*f$ is homogeneous of degree $-k$.

3.a. there exists a constant elasticity of product-(factor-) substitution along the boundary of $f(x)$, equal to σ;[1]
  b. there exists a constant elasticity of output price-(factor price-) substitution along the boundary of $*f(x)$, equal to $1/\sigma$.

Proof: (1) and (2) follow from properties 10.2.3 and 10.2.4.3. Let $\bar{f}(x) :=$ Bnd $f(x)$, the boundary of $f(x)$. As there exists a constant elasticity of product substitution, $\bar{f}(x)$ may be represented by the CES-function as follows: for each x, there exists a scalar η such that

$$\bar{f}(x) = \left\{ y \,\middle|\, \left(\sum_{i=1}^{m} \delta_i y_i^{\rho} \right)^{1/\rho} = \eta \right\}, \quad \text{with } \sigma = 1/(1-\rho).$$

Uzawa (1964) has shown that

$$*\bar{f}(x) = \left\{ q \,\middle|\, \left(\sum_{i=1}^{m} (q_i/\delta_i)^{\bar{\rho}} \right)^{1/\bar{\rho}} = 1/\eta \right\}, \qquad \text{with } \bar{\rho} = \rho/(\rho-1).$$

The elasticity of substitution of this set is $\bar{\sigma} = 1/(1-\bar{\rho}) = (\rho-1) = 1/\sigma$. □

The boundary of an output-set given some input x has been called the *transformation curve* corresponding to the quantity x, Bnd $f(x)$. Similar definitions are applicable for the *output price-isoquant*, Bnd $*f(x)$, and the *input price-isoquant* Bnd, $*f^{-1}(y)$.

1. The elasticity of substitution between any two factors x_1 and x_2 is defined by

$$\sigma := \frac{d(x_1/x_2)}{(x_1/x_2)} \bigg/ \frac{d(dx_1/dx_2)}{(dx_1/dx_2)} .$$

Fig. 3.4.1. Isoquants with constant elasticity of substitution and the corresponding input price-isoquants.

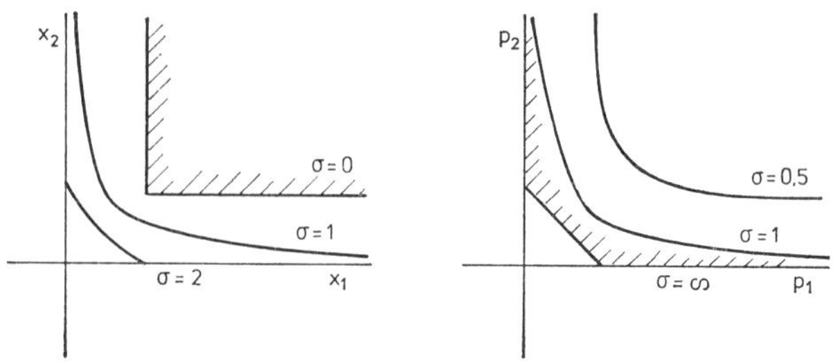

An illustration of property 3.4.1.2 is provided by the input-output multifunction $f(x_1, x_2) = \{y | 0 \leqq y \leqq x_1^\alpha x_2^\beta\}$ in section 3.3; here it was shown that $^*f(x_1, x_2) = \{q | q \leqq x_1^{-\alpha} x_2^{-\beta}\}$.

Illustrations of property 3.4.1.3 are given in fig. 3.4.1 for isoquants. If no substitution between input is possible ($\sigma = 0$), the technology reduces to the Leontief technology with fixed coefficients; in this case, there is perfect substitution ($\sigma = \infty$) between input prices. The reasoning may also be reversed, of course, so that a linear isoquant technology ($\sigma = \infty$) implies a unique relative price composition ($\sigma = 0$).

If the substitution between inputs is constantly equal to 1, the model reduces to the Cobb-Douglas technology; in this case the substitution between optimal input prices corresponding to a given output is also equal to 1.

Intermediate cases are represented by the C.E.S. technology.

Fig. 3.4.2. Transformation curves with corresponding output price-isoquants

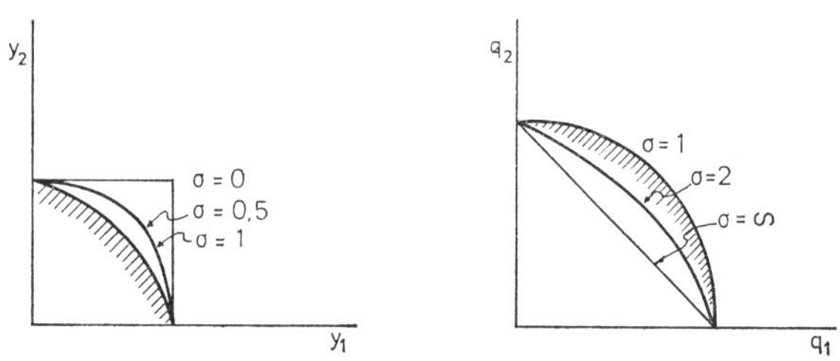

A similar reasoning can be given for the output sets $f(x)$ and the output price-sets ${}^{*}_{-}f(x)$ corresponding to a given input x (see fig. 3.4.2).

The following property of the input-output price, resp. output-input price multifunction, is deduced from properties 10.5.1 and 9.2.3.4, resp. 9.2.2.4: (see also section 10.5).

Property 3.4.2

Let the input-output multifunction $f: R^m \to R^n$ be a convex star process. Then:

1. the input-output price multifunction ${}^{*}_{-}f: R^m \to R^{n*}$ is point-starred, quasi-convex, graph-closed and l.h.c.;
2. the output-input price multifunction ${}^{*}_{+}(f^{-1}): R^n \to R^{m*}$ is point-aureoled, quasi-convex, graph-closed and l.h.c.

These results are complementary to those obtained by Shephard (1970, chapter 10).

Next, the effects of the adjoint operation on input-output multifunctions will be considered. The adjoint operation generates a relation between output prices *and* input prices such that, if the technology permits, the excess value of output over input (or profit) equals one. The following proposition is deduced from properties 10.3.1.6, 10.3.1.5, 10.5.1.4, 10.5.2.4 and 10.5.2.3.

Property 3.4.3

Let $f: R^m \to R^n$ be an input-output multifunction. Then:

1. the corresponding price-cost multifunction $f^{*}_{+}: R^{n*} \to R^{m*}$, and its inverse f^{*-1}_{+}, are both convex processes.
2. if the input-output multifunction, f, is starred, then the price-cost multifunction, f^{*}_{+}, and the output-input multifunction, f^{-1}, are aureoled, and the cost-price multifunction f^{*-1}_{+} is starred again.

The polarity operation may be applied to the price-cost multifunctions; and properties 3.4.1 and 3.4.2 are valid for the corresponding input-output multifunction. If the adjoint operation is applied to a price-cost multifunction, one obtains an input-output multifunction which is a convex process.

The various multifunctions describing the technology and defined in section 3.3 are related as is shown by the diagram in fig. 3.4.3.

Fig. 3.4.3. Representations of a technology

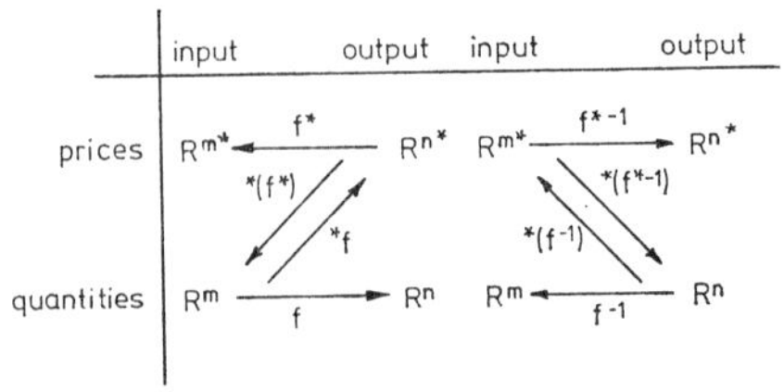

If it is known that the technology is closed and convex, then each of the multifunctions f, f^{-1}, f_+^* or f_+^{*-1} gives complete information about the technology and, given one, the others can be derived.

If it is also known that the technology is starred, then each of the just-mentioned multifunctions can also be derived from any of the four polar multifunctions (by virtue of property 9.2.2.5 and 9.2.3.5).

This means that one can estimate the technology in the space in which most information is available and derive the properties of it in the other spaces.

3.5. PREFERENCE ORDERINGS OF INPUTS

The last two sections of this chapter will deal with the input-side of a production technology. In this section it will be shown that, under certain assumptions, a producer has a preference ordering of all combinations of input quantities and all combinations of input prices. These preferences, dependent of course on the production technology, determine the producer's demand for inputs.

On the other hand, it is possible under some conditions to derive from the behavior of the producer (as expressed by his demand) his preferences for inputs. And with these preferences, the input-side of the technology can be calculated. The assumptions necessary to derive from the (observable) demand-multifunctions, the (unknown) production-technologies, are given in the next section. The main properties in both sections are based on the work by Weddepohl (1970) for a consumption model, but they can be transposed to any choice model, such as the production model outlined in this chapter.

Suppose that a producer has a technology that obeys the following assumptions:

Assumption 3.5.1 (*on the technology*)

The input-output multifunction $f: X \to Y$ is a non-negative, monotonously increasing, convex process.

Assumption 3.5.2 (*on non-satiation*)

Every non-negative scalar multiple of output $\bar{y}$ can be produced, i.e. $\forall \lambda \geqq 0$, $\exists x \in X: \lambda \bar{y} \in f(x)$.

From the technology $f: X \to Y$, a sub-technology can be derived for a given composition $\bar{y}$ of an output bundle. This sub-technology indicates the levels of output of composition $\bar{y}$, and is defined on the subset $X_{\bar{y}}$ of X enabling the producer to produce any positive level λ of $\bar{y}$:

$$X_{\bar{y}} := \cup \{f^{-1}(\lambda \bar{y}) \mid \lambda > 0\}.$$

The sub-technology for $\bar{y}$ is determined by the *production-level function* $f_{\bar{y}}: X_{\bar{y}} \to R_{++}$, defined by

$$f_{\bar{y}}(x) := \max \{\lambda \mid \lambda \bar{y} \in f(x)\}.$$

Property 3.5.1.

If assumptions 3.5.1 and 3.5.2 are satisfied for the input-output multifunction f, then $f_{\bar{y}}$ is a non-negative, monotone increasing, quasi-concave and continuous function.

Proof

Since the graph of the less-closure of $f_{\bar{y}}$, the set $\{(x, \lambda \bar{y}) \mid \lambda \bar{y} \in f(x)\}$, is equal to the intersection of a (convex) subspace and the (convex) graph of f, it is also a convex set (see property 8.3.3). This is sufficient for $f_{\bar{y}}$ being quasi-concave (see section 10.2). Via a similar reasoning it follows that $f_{\bar{y}}$ is non-negative and monotone increasing. Since the less-closure of $f_{\bar{y}}$ is lower hemi-continuous (see property 10.3.1.4), it follows that the function $f_{\bar{y}}$ is continuous (see section 8.2). □

The inverse of the production-level function, $f_{\bar{y}}^{-1}(\lambda)$, determines the input sets (see figs. 3.3.2 or 3.4.1). The boundaries of the input sets are the isoquants in $X_{\bar{y}}$. The isoquants contain all inputs enabling the producer to produce a given level of output, and between which the producer is therefore indifferent.

Thus, the set $X_{\bar{y}}$ of inputs relative to $\bar{y}$ can be ordered according to the output level obtained. The *preference relation* $\lesssim$ of producers on $X_{\bar{y}}$ is defined by:

$$x_1 \lesssim x_2 \Leftrightarrow f_{\bar{y}}(x_1) \leqq f_{\bar{y}}(x_2).$$

Property 3.5.2.

The production-level function $f_{\bar{y}} : X_{\bar{y}} \to R_{++}$ is non-negative, monotone increasing, quasi-concave and continuous, if and only if the preference relation $\lesssim$ on $X_{\bar{y}} \subseteq R^m_+$ satisfies the following conditions:

1. transitive : $x_1 \lesssim x_2$ and $x_2 \lesssim x_3 \Rightarrow x_1 \lesssim x_3$.
2. complete : $\forall x_1, x_2 \in X_{\bar{y}} : x_1 \lesssim x_2$ or $x_2 \lesssim x_1$.
3. monotone : $x_1 \leqq x_2 \Rightarrow x_1 \lesssim x_2$ and
   $x_1 < x_2 \Rightarrow x_1 \prec x_2$.
4. convex : $\forall x_1 \in X_{\bar{y}}$, the set $\{x | x_1 \lesssim x\}$ is convex.
5. continuous : $x_1 \lesssim x_2$ and $x_2 \lesssim x_3 \Rightarrow \exists \alpha \in [0, 1]$:
   $x_2 \sim (\alpha x_1 + (1-\alpha) x_3)$.

Proof
Let $f_{\bar{y}}$ be given. Then $X_{\bar{y}} \subseteq \text{Dom} f_{\bar{y}} = R^m_+$, and both transitivity and completeness follow from the definition of $\lesssim$. Since for each $\lambda > 0$, the input set $f^{-1}_{\bar{y}}(\lambda)$ is closed in $X_{\bar{y}}$, and is convex and monotone for $\leqq$, the other properties follow. The converse statement has been shown by Debreu (1959, section 4.6). □

Next, assume that there exist prices for inputs and that the amount which the producer can spend on inputs is equal to 1. The choice sets of the producer are then determined by the prevailing prices:

$$M(p) = \{x \in X_{\bar{y}} | px \leqq 1\}.$$

The producer will 'of course' choose the maximal elements according to $\lesssim$ in $M(p)$, with which input he can obtain the highest level of output. This, however, is an assumption which has to be explicitly stated (see assumption 3.7.1 below). Another problem is whether there exist maximal elements in the producer's choice sets; in order to answer this question, we must look at the price structure corresponding to the technology.

In section 3.3 the input price set $\overset{*}{_+}f^{-1}(\bar{y})$ is defined which corresponds with the input set $f^{-1}(\bar{y})$. The input price set contains all prices that give a value of at least 1 to any element of the input set. In a similar way, the polar set of the input set $f^{-1}_{\bar{y}}(\lambda)$ can be derived. The set of all input prices that enable the producer to buy an input with which he can produce a non-zero level of production of composition $\bar{y}$, is called the *set of input prices relative to* $\bar{y}$ and is denoted by $P_{\bar{y}}$:

$$\begin{aligned} P_{\bar{y}} := & \cup \{[f_{\bar{y}}^{-1}(\lambda)]_{+}^{*} | \lambda > 0\} \\ = & \cup \{{}_{+}^{*}f^{-1}(\lambda\bar{y}) | \lambda > 0\} \\ = & \{p \in R^{m*} | \exists \lambda > 0 : M(p) \cap \operatorname{Int} f^{-1}(\lambda\bar{y}) = \phi\}. \end{aligned}$$

Fig. 3.5.1. Isoquants in $X_{\bar{y}}$ and their corresponding isoquants in $P_{\bar{y}}$

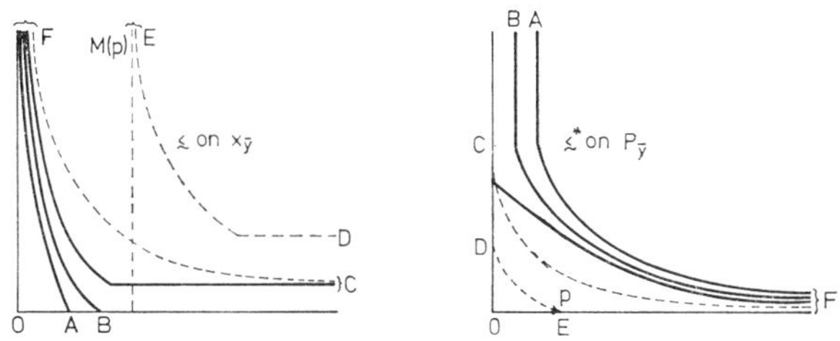

This equality follows from the definition of the dual space given in section 9.1, as the dimension of $f^{-1}(\lambda\bar{y})$ is equal to m, the dimension of $M(p)$. This way of writing makes clear that it is very well possible that for some $p \in P_{\bar{y}}$, there exists no $\lambda > 0$ such that $M(p) \cap \operatorname{Int} f_{\bar{y}}^{-1}(\lambda) = \phi$ and $M(p) \cap f_{\bar{y}}^{-1}(\lambda) \neq \phi$, implying that no maximal element exists in $M(p)$. See, for example, $M(p)$ in fig. 3.5.1. This case however will be excluded by means of suitable assumptions, given in section 3.6.

Just as the isoquants in $X_{\bar{y}}$ generate a preference relation $\lesssim$ in $X_{\bar{y}}$, which is equivalently expressed as:

$$x^1 \lesssim x^2 \Leftrightarrow \exists \lambda > 0 : x^1 \in \operatorname{Bnd} f_{\bar{y}}^{-1}(\lambda) \quad \text{and} \quad x^2 \in f_{\bar{y}}^{-1}(\lambda),$$

so the isoquants in $P_{\bar{y}}$ generate a relation $\lesssim^*$ in $P_{\bar{y}}$. A price p^2 is said to be *as least as costly* as p^1, or $p^1 \lesssim^* p^2$, if, given the amount to be spend, for every bundle x_2^2 in the choice set $M(p^2)$, there exist a bundle x^1 in $M(p^1)$ such that x^1 is preferred to x^2, i.e.: $\forall x^2 \in M(p^2), \exists x^1 \in M(p^1)$: $x^1 \gtrsim x^2$. This relation $\lesssim^*$ on $P_{\bar{y}}$ is called a *costing relation* on the prices in $P_{\bar{y}}$ and is equivalently expressed as:

$$p^1 \lesssim^* p^2 \Leftrightarrow \exists \lambda > 0 : p^1 \in \operatorname{Bnd}^* f_{\bar{y}}^{-1}(\lambda) \quad \text{and} \quad p^2 \in {}^* f_{\bar{y}}^{-1}(\lambda).$$

This relation has been defined by Milleron (1968) and, independently, by Weddepohl (1970). The equivalent expression given above seems, however, more neutral if the choice sets are subsets of $P_{\bar{y}}$ rather than of $X_{\bar{y}}$, as in the case of public goods.

The monotonicity assumption in 3.5.1 implies that a larger price for $\leqq$ is always a dearer price for $\lesssim^*$. The inverse statement, of course, is not true. Suppose, for example, that a certain amount of money must be spent, either in a supermarket with prices p^1, or in a supermarket with prices p^2; then p^1 is costlier for the buyer than p^2 if a better basket can be bought by him at prices p^2 than at prices p^1. Without extra assumptions, the costing relation $\lesssim^*$ on $P_{\bar{y}}$ does not have the same properties as the relation $\lesssim$ on $X_{\bar{y}}$; e.g. if in fig. 3.5.1 both the isoquants in $X_{\bar{y}}$ converging in C are feasible, then there exists a price (C) in $P_{\bar{y}}$ which is equivalent to prices on two different isoquants. This clearly contradicts the transitivity condition. In order to derive and valuate what is a necessary assumption, the concepts of satiation and dispensability will be defined.

3.6. SATIATION FOR AND DISPENSABILITY OF INPUTS

An input-component i is said to be *indispensable* (or: *essential*) for the production of $\bar{y}$ via f, if $x_i = 0$ implies that $x := (x_1, \ldots, x_i, \ldots, x_l) \notin X_{\bar{y}}$. (See fig. 3.5.2.)

Fig. 3.5.2. Input x_1 is indispensable (essential) to $\bar{y}$

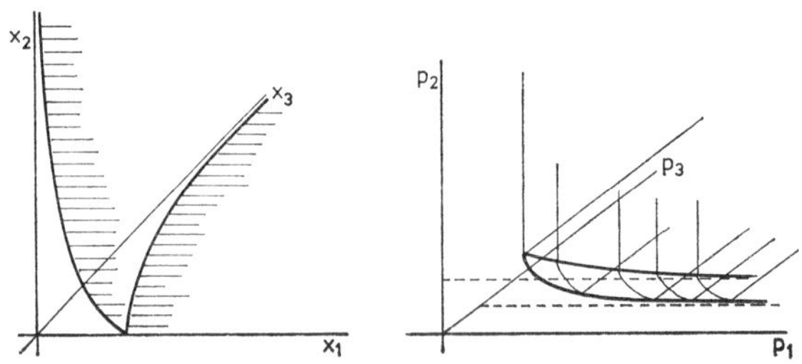

If a positive level α can be indicated, which is indispensable in the production of $\bar{y}$, i.e. $x_i = \alpha \Rightarrow x \notin X_{\bar{y}}$, then we can assume that any level between α and 0 is also indispensable. In this case, the input-set $X_{\bar{y}}$ is transposed by withdrawing the vector $\alpha e_i := (0, \ldots, \alpha, \ldots, 0)$, with α on the i-th place, from $X_{\bar{y}}$. The resulting input set $X'_{\bar{y}} = X_{\bar{y}} - \{\alpha e_i\}$ has only one (zero-) level of indispensability (see assumption 3.6.2).

It follows from the definition that component i is *dispensable* in the production of $\bar{y}$, if there exists an $x \in X_{\bar{y}}$ such that $x_i = 0$. As there exists a $p \in P_{\bar{y}}$ such that $px = 1$ and $(p+e_i)x = 1$ whenever $x_i = 0$, it is also true that component i is dispensable if and only if there exists a $p \in P_{\bar{y}}$ such that $p+e_i \sim^* p$. This, again, is equivalent to saying that component i is indispensable if, and only if, for all $p \in P_{\bar{y}}$, $p+e_i \prec^* p$. (See property 3.6.1.)

On the other hand, the production of $\bar{y}$ is said to be *insatiable* for component j of input x, if for all $x \in X_{\bar{y}}$: $x+e_j \succ x$.

This means that an increase of component j in the input always strictly increases the quantity of output (see fig. 3.6.1).

Fig. 3.6.1. The production of $\bar{y}$ is insatiable for input x_1

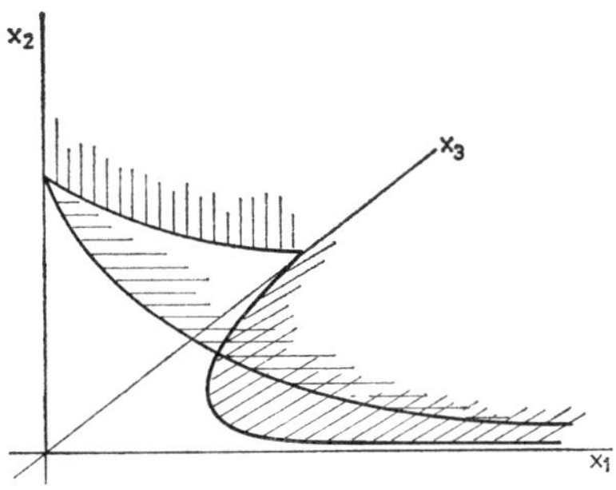

The production of $\bar{y}$ is satiable for component i of the input x, if there exists an $\bar{x} \in X_{\bar{y}}$ such that $\bar{x}+e_i \sim \bar{x}$. If the production of $\bar{y}$ is satiable for all components of input x simultaneously, i.e. $\exists \bar{x} \in X_{\bar{y}} : \bar{x}+\Sigma e_i \sim \bar{x}$, then there exist (given assumption 3.5.1) a maximum level $\bar{\lambda}$ for the production of $\bar{y}$. For at that input $\bar{x}$, $\bar{\lambda} = f_{\bar{y}}(\bar{x}) = f_{\bar{y}}(\bar{x}+\mu e)$, for all $\mu \geqq 0$. In this case, the set of inputs relative to $\bar{y}$ has a thick indifference class (i.e. an indifference class with a non-empty interior in $X_{\bar{y}}$), viz. the satiation class. Satiation in the production of $\bar{y}$ is ruled out by assumption 3.5.2.

It should be noted that the case in which all components of input x are simultaneously dispensable for the production of $\bar{y}$ is ruled out by assumption 3.5.1, for $x = 0$ implies that $f(x)$ contains no element y having a positive component.

An analogous reasoning can be applied to the case of indispensable commodities. This results in the following property:

Property 3.6.1.

Given the preference orderings $\precsim$ on $X_{\bar{y}}$ and $\precsim^*$ on $P_{\bar{y}}$, the characteristics of a commodity *i* defined above are expressed equivalently in the following way:

1. indispensable : $x_i = 0 \Rightarrow x \notin X_{\bar{y}}$;
2. insatiable : $p_i = 0 \Rightarrow p \notin P_{\bar{y}}$;
3. dispensable : $\exists p \in P\bar{y} : p \sim^* p + e_i$;
4. satiable : $\exists x \in X_{\bar{y}} : x \sim x + e_i$.

The following assumption 3.6.1 restricts the class of production technologies determined by assumption 3.5.1.

Assumption 3.6.1. (*On satiation and dispensability*)

The technology represented by $f_{\bar{y}} : X_{\bar{y}} \rightarrow R_+$ and determined by assumption 3.5.1 has the following properties:

1. If production is satiable for some input-component *i*, then $\forall x \in X_{\bar{y}}$, $\exists \lambda \geqq 0 : x + \lambda e_i \sim x + \lambda e_i + e_i$;
2. If some input-component *i* is dispensable for production, then $\forall p \in P_{\bar{y}}, \exists \lambda \geqq 0 : p + \lambda e_i \sim^* p + \lambda e_i + e_i$.

This assumption does not imply that production is satiable for some or all components of input. E.g., the Cobb-Douglas function, which represents a technology that is insatiable for all components, is not ruled out by this assumption (in this case the set $X_{\bar{y}}$ of inputs is open in R^m). The assumption requires, however, that if production is satiable somewhere for some component, then increasing the quantity of this input-component from any point in $X_{\bar{y}}$ will ultimately cause no further increase in output. Since this 'ultimate' point $(x + \lambda e_i)$ may be arbitrary far from x, it is not a strong assumption at all: if the inputs are narrowly enough specified, then it is always true.

The assumption implies that (1) if the price of an input-component is zero, its demand can never be infinite, and (2) if an input-component is dispensable, then its price will not be infinite. Some C.E.S. functions are therefore ruled out by this assumption (see also section 4.1). This assumption is essential for the following two important properties 3.6.2 and 3.7.1.

Property 3.6.2.

Let the technology satisfy the assumptions 3.5.1 and 3.6.1. Then the costing relation $\precsim^*$ on $P_{\bar{y}}$ is a complete preorder.

Proof
The reflexivity of $\precsim^*$ follows from Bnd ${}^*_+f_{\bar{y}}^{-1}(\lambda) \subseteq {}^*_+f_{\bar{y}}^{-1}(\lambda)$, such that $p \precsim^* p$.
The transitivity of $\precsim^*$ is shown as follows.

Fig. 3.6.2.

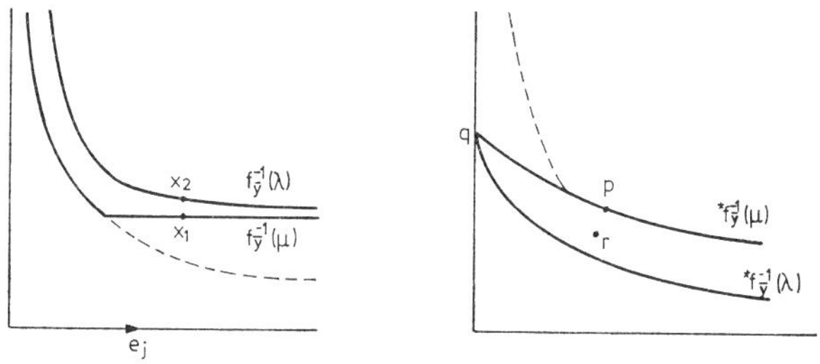

Suppose that there exist positive λ, μ, $\lambda \neq \mu$, such that $p \in$ Bnd ${}^*_+f_{\bar{y}}^{-1}(\mu)$, $q \in {}^*_+f_{\bar{y}}^{-1}(\mu) \cap$ Bnd ${}^*_+f_{\bar{y}}^{-1}(\lambda)$, and $r \in {}^*_+f_{\bar{y}}^{-1}(\lambda)$. If $r \in {}^*_+f_{\bar{y}}^{-1}(\mu)$, then $p \precsim^* q \precsim^* r$, implying transitivity of $\precsim^*$. Suppose that $r \notin {}^*_+f_{\bar{y}}^{-1}(\mu)$, as in fig. 3.6.2, then $p \succ^* r$ and $q \in$ Bnd ${}^*_+f_{\bar{y}}^{-1}(\mu) \cap$ Bnd ${}^*_+f_{\bar{y}}^{-1}(\lambda)$, for $\lambda \neq \mu$.

It follows that there exists some component q_j in q which is zero, and that the budget set $M(q)$ is such that $M(q) \cap f_{\bar{y}}^{-1}(\mu) \neq \phi$ and $M(q) \cap f^{-1}(\lambda) = \phi$ (see fig. 3.6.2). Therefore, for some $x_1 \in$ Bnd $f_{\bar{y}}^{-1}(\mu)$ it is true that $x_1 + e_j \in$ Bnd $f_{\bar{y}}^{-1}(\mu)$, or $x_1 \sim x_1 + e_j$, and for another isoquant it is so that there does not exist a $\lambda \geqq 0$ such that for some x, $(x + \lambda e_j) =: x_2 \in$ Bnd $f_{\bar{y}}^{-1}(\lambda)$ implies that $x_2 + e_j \in$ Bnd $f_{\bar{y}}^{-1}(\lambda)$, or $x + \lambda e_j \sim x + \lambda e_j + e_j$. This contradicts assumption 3.6.1.

Completeness of $\precsim^*$ is implied by the definition of $P_{\bar{y}}$, by which for every $p \in P_{\bar{y}}$, there exists a $\lambda > 0$ such that $p \in {}^*_+f_{\bar{y}}^{-1}(\lambda)$. □

It finally should be noted that assumption 3.6.1 in property 3.6.2 can be replaced by the assumption that $X_{\bar{y}}$ is compact.

3.7. THE DEMAND MULTIFUNCTION AND THE PRICE MULTIFUNCTION

If the choice sets in $X_{\bar{y}}$ generated by prices in $P_{\bar{y}}$ always contain maximal elements, then a set of maximal elements can be associated with every price. If one assumes that all agents will actually choose or demand the bundle which is best for them, then the demand multifunction can be sensibly defined.

The necessary condition mentioned above can be derived from the assumptions 3.5.1 and 3.6.1:

Property 3.7.1.

Let the technology be given by assumptions 3.5.1 and 3.6.1. Then:

1. All choice sets $M(p)$ in $X_{\bar{y}}$, generated by $p \in P_{\bar{y}}$, have maximal (or most preferred) elements for $\lesssim$, i.e.

$$P_{\bar{y}} = \{p \mid \exists \lambda > 0: M(p) \cap \operatorname{Int} f_{\bar{y}}^{-1}(\lambda) = \phi \text{ and } M(p) \cap f_{\bar{y}}^{-1}(\lambda) \neq \phi\}.$$

2. All choice sets $M(x)$ in $P_{\bar{y}}$, generated by $x \in X_{\bar{y}}$, have maximal (or most costly) elements for $\lesssim^*$, i.e.

$$X_{\bar{y}} = \{x \mid \exists \lambda > 0: M(x) \cap \operatorname{Int} {}^*f_{\bar{y}}^{-1}(\lambda) = \phi \text{ and } M(x) \cap {}^*f_{\bar{y}}^{-1}(\lambda) \neq \phi\}.$$

Proof

1. A choice set $M(p)$ in $X_{\bar{y}}$ has, by definition, a maximal element $\bar{x}$ if $\bar{x} \in M(p)$ and $x \succ \bar{x} \Rightarrow x \notin M(p)$, i.e., there exists a $\lambda > 0$ such that $\bar{x} \in M(p) \cap f_{\bar{y}}^{-1}(\lambda)$ and $\operatorname{Int} f_{\bar{y}}^{-1}(\lambda) \cap M(p) = \phi$. This must be true for all $p \in P_{\bar{y}}$.

If $p > 0$, then $p \in \operatorname{Int} P_{\bar{y}}$ and $M(p) \cap X_{\bar{y}}$ is a bounded set. As $M(p)$ is also closed in $X_{\bar{y}}$, it has a maximal element.

If $p \in P_{\bar{y}}$ and $p \not> 0$, then by the definition given of $P_{\bar{y}}$, there exists a $\lambda > 0$ such that $p \in \operatorname{Bnd} {}^*f_{\bar{y}}^{-1}(\lambda)$, or $M(p) \cap \operatorname{Int} f_{\bar{y}}^{-1}(\lambda) = \phi$ and a component j in p exists for which $p_j = 0$. Therefore, production is satiable for input component j and (from assumption 3.6.1) $\forall x \in X_{\bar{y}}$, $\exists \lambda \geqq 0$: $x + \lambda e_j \sim x + \lambda e_j + e_j$; by which there exists an $\bar{x} \in \operatorname{Bnd} f_{\bar{y}}^{-1}(\lambda)$, such that $\bar{x} \sim \bar{x} + e_j$. Since $p \in \operatorname{Bnd} {}^*f_{\bar{y}}^{-1}(\lambda)$, $\bar{x} \in M(p) \cap f_{\bar{y}}^{-1}(\lambda)$.

2. An analogous reasoning can be applied to the polar case. As the set $X_{\bar{y}}$ is defined in another way than the set $P_{\bar{y}}$, it may be noted that

$$\begin{aligned} X_{\bar{y}} &:= \cup \{f_{\bar{y}}^{-1}(\lambda) \mid \lambda > 0\} = \cup \{[{}^*f_{\bar{y}}^{-1}(\lambda)]^* \mid \lambda > 0\} \\ &= \{x \mid \exists \lambda > 0: M(x) \cap \operatorname{Int} {}^*f_{\bar{y}}^{-1}(\lambda) = \phi\}, \end{aligned}$$

according to property 10.5.2. □

Next we assume that the behavior of agents is in accordance with their opinions or ambitions:

Assumption 3.7.1. (on behavior)

Each agent in the economy chooses a maximal element in his choice set.

This assumption is natural if a maximal element is a best element for $\lesssim$ in the quantity space; it is less natural, however, if it is a most costly element for $\lesssim^*$ in the price space. In this case, special consideration is needed to interprete this assumption.

The multifunction which associates with each price in $P_{\bar{y}}$ the set of maximal elements in the choice set generated by that price is called the *demand multifunction* $h: P_{\bar{y}} \to X_{\bar{y}}$,

$$\begin{aligned} h(p) &:= \{x \in X_{\bar{y}} | x \text{ is a maximal element of } M(p) \text{ for } \lesssim\} \\ &= \{M(p) \cap f_{\bar{y}}^{-1}(\lambda) | p \in {}^*f_{\bar{y}}^{-1}(\lambda)\} \end{aligned}$$

The last equality is checked as follows: if $\bar{x}$ were not maximal, then $\bar{x} \succ x$ and $p\bar{x} \leqq 1$, or $x \in \operatorname{Int} f_{\bar{y}}^{-1}(\lambda)$ for λ such that $p \in {}^*f_{\bar{y}}^{-1}(\lambda)$; by definition $M(p) \cap \operatorname{Int} f_{\bar{y}}^{-1}(\lambda) = \phi$. If $\bar{x} \notin f_{\bar{y}}^{-1}(\lambda)$, then $p\bar{x} < 1$, and $\bar{x}$ is not maximal in $M(p)$.

The inverse $h^{-1}: X_{\bar{y}} \to P_{\bar{y}}$ is said to be the *price multifunction.* It follows from the equivalent expressions above that:

$$\begin{aligned} h^{-1}(x) &:= \{p \in P_{\bar{y}} | x \in h(p)\} \\ &= \{M(x) \cap {}^*_+f_{\bar{y}}^{-1}(\lambda) | x \in f_{\bar{y}}^{-1}(\lambda)\} \\ &= \{p \in P_{\bar{y}} | p \text{ is a maximal element of } M(x) \text{ for } \lesssim^*\}. \end{aligned}$$

The price multifunction, therefore, associates with each bundle of commodities the (set of) prices at which the bundle will be demanded. This price, however, is also the most costly price for the agent in the choice set $M(x)$ in $P_{\bar{y}}$. The following property applies to both multifunctions, h and h^{-1}:

Property 3.7.2.

Given assumptions 3.5.1, 3.6.1 and 3.7.1, both the demand multifunction $h: P_{\bar{y}} \to X_{\bar{y}}$ and the price multifunction $h^{-1}: X_{\bar{y}} \to P_{\bar{y}}$ are non-empty, point-closed and point-convex. The value px of each element in the images is equal to 1, i.e., $p \cdot h(p) = h^{-1}(x) \cdot x = 1$.

Proof

By 3.5.1 and 3.6.1, for any $p \in {}^*_+f_{\bar{y}}^{-1}(\lambda) \subseteq P_{\bar{y}}$, the sets $M(p) \cap f_{\bar{y}}^{-1}(\lambda)$ and $M(x) \cap {}^*_+f_{\bar{y}}^{-1}(\lambda)$ are non-empty (see property 3.7.1). As all sets $M(p)$, $M(x)$, $f_{\bar{y}}^{-1}(\lambda)$ and ${}^*_+f_{\bar{y}}^{-1}(\lambda)$ are closed and convex, so is their intersection.

Further, $p \in M(x)$ implies that $px \leqq 1$ and $p \in {}^*_+f_{\bar{y}}^{-1}(\lambda)$ implies that $px \geqq 1$. □

The last property mentioned here and induced by assumptions 3.5.1, 3.6.1 and 3.7.1, is that the agents will reveal their opinions about the elements in $X_{\bar{y}}$ or $P_{\bar{y}}$ in a consistent way. An input $x \in X_{\bar{y}}$ is said to be *revealed preferred* to $y \in X_{\bar{y}}$, if there exists a choice set $M(p)$ which contains y, but from which x (possibly equal to y) is chosen, i.e., $h^{-1}(x) \cap M(y) \neq \phi$. This concept was introduced by Samuelson (1938), who considered this binary relation as being the only binary relation which is empirically verifiable and therefore meaningful. The consistency or 'rationality' requirement relative to this relation : the weak axiom of revealed preference, can be deduced from the assumptions as a property: $h^{-1}(x) \cap M(y) \neq \phi$ and $h^{-1}(x) \cap h^{-1}(y) = \phi$ imply $h^{-1}(y) \cap M(x) = \phi$, because $h^{-1}(x) \cap M(y) \neq \phi$ implies that $x \gtrsim y$ and $h^{-1}(x) \cap h^{-1}(y) = \phi$ implies that $x \not\lesssim y$; if $h^{-1}(y) \cap M(x)$ were not empty, then $y \gtrsim x$, which contradicts the first statement.

Analogously, a binary relation revealing costing can be defined on the set of prices $P_{\bar{y}}$ (compare Weddepohl, 1970, p. 38). A price $p \in P_{\bar{y}}$ is said to be *revealed costlier* than $q \in P_{\bar{y}}$, if there exists a choice set $M(x)$ which contains q but from which p (possibly equal to q) is chosen, i.e., $h(p) \cap M(q) \neq \phi$. Again, properties of 'revealed costing' can be deduced.

The demand and price multifunctions, and all properties given above, are derived from the preference relations $\lesssim$ and $\lesssim^*$ given in property 3.5.1, which relations are induced again by the assumptions 3.5.1, 3.6.1 and 3.7.1.

The reverse problem is just as interesting: is it possible to generate the preference relation $\lesssim$ or the costing relation $\lesssim^*$ (i.e. the unknown technology) from the (observable) demand or price multifunctions?

This question has been studied by many scholars (see Chipman e.a., 1971), most of whom are concerned with determining the 'integrability conditions' of the demand function. It has been shown that it is possible to generate a preference relation $\lesssim$ on $X_{\bar{y}}$ or $\lesssim^*$ on $P_{\bar{y}}$ with the properties given in 3.5.1, if one assumes that the demand or price multifunctions obey the properties of this section (3.7.1 and 3.7.2, the weak and strong axiom of revealed preference), analogues to assumptions 3.6.1 and 3.7.1 (satiation and behavior) and, in addition, a generalized Lipschitz condition for h and h^{-1} to guarantee weak continuity (see Weddepohl, 1970, p. 134). The question can be theoretically answered for many cases; for these cases, 'only' empirical problems of investigation remain to be undertaken.

4. Consumption and production of public goods

4.1. THE CONSUMPTION MODEL

Just as a production activity, a consumption activity can be considered as a process with an input and an output. The analogue to the production-level function is called the *utility function* $u : X \rightarrow R_{++}$, which assigns to each vector of quantities consumed a level of satisfaction. The consumption model is a special case of the production model, because only the one-dimensional utility function is considered. The corresponding properties given in the previous chapter can be directly applied on consumption.

In particular, with each utility function a binary preference relation can be associated which is, under certain conditions, an ordering of the set of inputs (see sections 3.5–3.7). According to property 3.5.2, the consumer's taste or 'technology' can equivalently be described by means of a utility function or a *preference relation* $\lesssim$ on $X \subseteq R^m_+$, for some dimension m:

Property 4.1.1.

Each consumer's taste can be described by a non-negative, monotone increasing, quasi-concave and continuous utility function, if and only if each consumer has a preference relation $\lesssim$ on the set of consumption bundles which is a non-negative, monotone, convex, continuous and complete preorder.

It should be mentioned that non-negativity and monotonicity are not essential for the assumption or the property. This is introduced only to simplify subsequent arguments.

The preference relation $\lesssim$ on the consumption set X divides the consumption set in two parts, given some consumption bundle $x \in X$: the set of consumption bundles which are preferred to x or equivalent to x, and the set of bundles which are not preferred to x. The first set is called the *upper-preference set* and defined by:

$$C(x) := \{y \in X | x \lesssim y\}.$$

The boundary of an upper-preference set is called an *indifference-curve* and contains all commodity combinations from which the consumer derives the same level of satisfaction.

This set is defined for every $x \in X$ and defines thus a multifunction $C: X \to X$, called the *upper-preference multifunction*. The inverse of this multifunction is equal to

$$C^{-1}(x) = \{y \in X | y \lesssim x\}.$$

*Fig. 4.1.1. The preferred set at x, C(x), and its polar set, *C(x)*

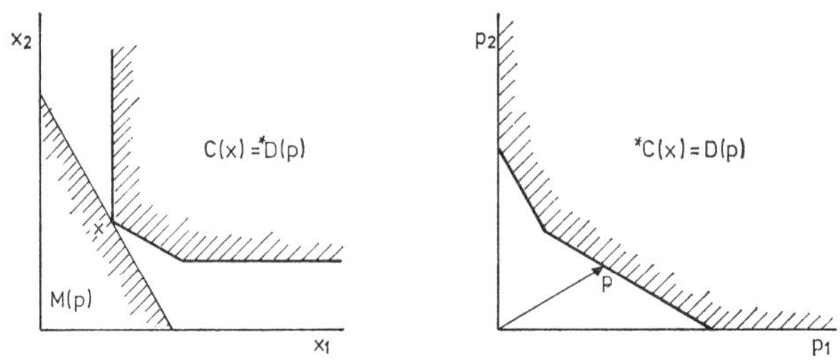

The representation of the taste (or 'technology') of a consumer through a preference multifunction has some advantages in this approach. Firstly because the maximal elements in a choice set $M(p)$ can be represented by the intersection of two sets (see fig. 4.1.1), since all elements not in $C(x)$ are preferred less than x. This representation is easily transposed in the dual space. Secondly, no reference at all is made to the level of utility, which is an artificial concept as utility is not a cardinal but an ordinal entity.

The following assumptions will be made about the preferences of consumers, in terms of the upper-preference multifunction $C: X \to X$.

Assumptions 4.1. (on consumers)

4.1.1. The consumption set X, the effective domain of C, is convex.
4.1.2. The consumption set X is preordered, i.e.
$x \in C(x)$ and $C[C(x)] \subseteq C(x)$. (See section 8.1).

4.1.3. The upper-preferred sets are convex, i.e.
C is quasi-concave.

4.1.4. The equivalence sets are thin, i.e.
$C(x) \cap C^{-1}(x) = \text{Bnd } C(x)$, for all x.

4.1.5. All choice sets in both spaces have maximal elements, i.e.,
$p \in \text{Bnd } {}^*C(x)$ implies that $M(p) \cap C(x) \neq \phi$ and
$y \in \text{Bnd } C(x)$ implies that $M(y) \cap {}^*C(x) \neq \phi$, for all $x \in X$.

4.1.6. The preferred sets are nonnegative and monotone, i.e., C is nonnegative and monotone decreasing.

Assumptions 4.1.1, 4.1.2 and 4.1.3 are standard assumptions in the literature; 4.1.4 implies that the preference sets are closed and that the consumer is not satiated. It will be shown that this assumption makes the preference multifunction lower hemi-continuous. Assumption 4.1.5 is new; it restricts the class of preference preorderings by excluding preference sets that asymptotically support choice sets, such as some C.E.S. functions. It permits however, preference preorderings which imply that some commodities are indispensable (or for which consumers are insatiable) so that the consumption set is not closed in R^m (see section 3.6). These kinds of preference preorderings are excluded if one assumes that the consumption set is closed, which is usually done.

Assumption 4.1.6 can be replaced by the condition that the upper-preference sets are aureoled and that $X \cap -X \subseteq \{0\}$. The assumption 4.1.6 is chosen here to simplify the arguments (and figures).

For each consumer, a *set of consumer's-prices* P in R^{m*} can be derived from the preference relation $\lesssim$ on X, or the upper-preference sets $C(x)$ in X. This is done in a similar way as the derivation of the set of input prices $P_{\bar{y}}$ in section 3.5:

$$\begin{aligned} P &:= \cup \{{}^*C(x) | x \in X\} \\ &= \{p \in R^{m*} | \exists x \in X \colon M(p) \cap \text{Int } C(x) = \phi\}. \end{aligned}$$

Again, just as the indifference curves in X generate the preference relation $\lesssim$ on X, so do they generate a *costing relation* $\lesssim^*$ on P. A price p^2 is said to be *at least as costly* as p^1, or $p^1 \lesssim^* p^2$, if – given the amount to be spend – there exists a bundle x in the choice set $M(p^1)$ which is preferred to every bundle in the choice set $M(p^2)$, i.e. $p^1 \in \text{Bnd } {}^*C(x)$ and $p^2 \in {}^*C(x)$.

Given a costing relation $\lesssim^*$ on P, the *upper-costing set* with respect to $p \in P$ is defined by (see fig. 4.1.1)

$$\begin{aligned} D(p) &:= \{q \in P | p \lesssim^* q\} \\ &= \{{}^*C(x) | p \in \text{Bnd } {}^*C(x)\}. \end{aligned}$$

The costing relation $\lesssim^*$ is thus a subjective relation, different for each consumer. If applied on private goods, as is done in this section, then the consumer will prefer the least costly price. Since $p \leqq q$ implies that $p \lesssim^* q$, this price is closest to zero. In another context, however, the costing relation $\lesssim^*$ will indicate priorities given to public goods (see section 5.2), and the consumer will indicate the highest priority (here the costliest price). It is for that context that the relation $\lesssim^*$ is introduced.

The definitions and properties given in sections 3.6 and 3.7 can directly be transposed to the consumer model. It also can be noticed that the multifunction $^*C : X \to P$ is derived from $C : X \to X$ through the polarity operation defined in section 10.1. These multifunctions have the following property:

Property 4.1.2. (continuity of preference multifunctions)

Let the consumer's upper-preference multifunction $C : X \to X$ meet assumptions 4.1. Then both C and its polar multifunction $^*C : X \to P$ are continuous on X and graph-closed.

Proof

The lower hemi-continuity of C is shown as follows. Choose $x \in X$, $y \in C(x)$ and an open set U containing y. Choose $\bar{y} \in C(x)$ such that $\bar{y} \in \text{Int } C(x)$. By assumption 4.1.4, $\bar{y} \notin C^{-1}(x)$ and therefore $x \notin C(\bar{y})$. Call $V := \text{Int } C^{-1}(\bar{y})$; V is open and contains x. Since for all $z \in V$: $\bar{y} \in C(z)$, it follows that $C(z) \cap U \neq \phi$ and that C is l.h.c. by property 8.2.2 (see fig. 4.1.2).

Fig. 4.1.2. C l.h.c., resp. not u.h.c.

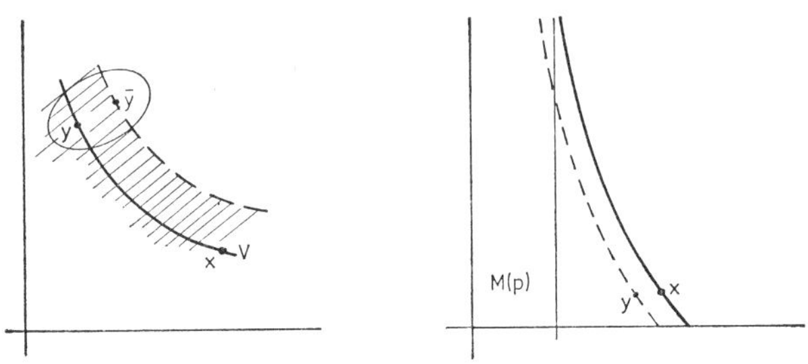

The upper hemi continuity of C follows from assumption 4.1.5: Choose $x \in X$ and a closed set M such that $M \cap C(x) = \phi$. Call $U := X \backslash M$; U is open and contains $C(x)$. As M and $C(x)$ are closed; they are strongly separated by a hyperplane, unless they are separated by an asymptotic support (see property 9.4.2.1 and fig. 4.1.2). In that case, $M(p)\cap C(x)=\phi$. for $p\in \text{Bnd } {}^*C(x)$. This case is excluded by assumption 4.1.5, implying that a y exists with $M(p)\cap C(y)\neq\phi$ but $M(p)\cap \text{Int } C(y)=\phi$.

Call $V:=\text{Int } C(y)$, then $x\in V$ and for all $z\in V: C(z)\subseteq U$. Therefore C is u.h.c.; since C is point-closed, C is also graph-closed according to property 8.2.2.1.

Analogous arguments can be applied to $^*C$. □

Property 4.1.3. (On the consumers' price space)

Let the consumer's preference multifunction meet assumptions 4.1. Then:

1. The price set P is convex and completely preordered.
2. The upper-costing multifunction $D: P\to P$ is quasi-concave, non-negative and monotone decreasing. A set of equivalent prices is thin, i.e. $D(p)\cap D^{-1}(p)=\text{Bnd } D(p)$, for all p.

Proof

1. Convexity of P follows from the convexity of a polar set $^*C(x)$ by property 9.2.2.4. The upper-costing multifunction D gives a relation on P which is reflexive, since $p\in D(p)$, and transitive, i.e. $D[D(p)]\subseteq D(p)$. This is shown as follows. Assume that there exists a $q\in D(p)$ such that $D(q)\nsubseteq D(p)$; cp. fig. 3.6.2. Then there exist x and y such that $p\in\text{Bnd } {}^*C(x)$, $q\in{}^*C(x)\cap\text{Bnd } {}^*C(y)$ and ${}^*C(y)\nsubseteq {}^*C(x)$. Thus $q\in\text{Bnd } {}^*C(x)\cap\text{Bnd } {}^*C(y)$ and $\text{Bnd } {}^*C(x)\neq\text{Bnd } {}^*C(y)$. Assumption 4.1.5 and the definition of a polar set imply that $M(q)\cap\text{Bnd } C(x)\neq\phi$ and $M(q)\cap\text{Bnd } C(y)\neq\phi$. Monotonicity (A4.1.6) and transitivity (A4.1.2) of C imply that $\text{Bnd } C(x)\cap\text{Bnd } C(y)\neq\phi$ and $\text{Bnd } C(x)=\text{Bnd } C(y)$. This contradicts $\text{Bnd } {}^*C(x)\neq\text{Bnd } {}^*C(y)$, by property 9.2.3.6. The relation is complete as $D^{-1}(p)\cap D(p)=P$.

2. Since $D^{-1}(p)=P\backslash\text{Int } D(p)$ and $D(p)$ is convex, it follows that $D^{-1}(p)$ is a concave set. Thus D is a quasi-concave multifunction by definition.

Since for every x, $\text{Cl Cone } C(x)=R^n_+$ and, by property 9.3.2, $0^+({}^*C(x))=\text{Cl Cone } {}^*C(x)=(R^n_+)^\circ_+=R^{n*}$, it follows that $^*C$ and therefore also D is non-negative and monotone decreasing.

Since $^*C$ is continuous, so are D and D^{-1}; therefore, $D(p)\cap D^{-1}(p)=\text{Bnd } D(p)$. □

Finally, the *demand multifunction* $h: P\to X$ for the consumer is defined by (see section 3.7 and fig. 4.1.1):

$$\begin{aligned} h(p) &:= \{x\in X \mid x \text{ is a maximal element of } M(p) \text{ for } \lesssim\} \\ &= \{M(p)\cap C(y) \mid y\in {}^*C^{-1}(p)\} = M(p)\cap {}^*D(p). \end{aligned}$$

In the same way, the consumer's *price multifunction* $h^{-1}: X \to P$ is equal to

$$h^{-1}(x) := \{p \in P \mid p \text{ is a maximal element of } M(x) \text{ for } \lesssim^*\} \\ = M(x) \cap {}^*C(x).$$

The consumer's demand- or price multifunctions give the elements which are equilibrium points for the consumer. They assign points which the consumer can obtain (in his choice set M) and wants to obtain (the best possible or maximal). This is expressed by the touching of two sets (see also fig. 4.1.1), one star shaped and the other aureole shaped: one determined by centripedal forces (the choice set) and the other by centrifugal forces (the preferred set).

Since $M(p)$, ${}^*D(p)$, $M(x)$ and ${}^*C(x)$ are graph-closed and point-convex multifunctions, the following property results from applying properties 9.2.2, 9.2.3 and 8.2.3 to the intersection of multifunctions:

Property 4.1.4. (*demand- and price multifunctions*)

Let the consumer's preference multifunction meet assumptions 4.1. Then the demand multifunction $h: P \to X$ and the price multifunction $h^{-1}: X \to P$ are non-empty, graph-closed and point-convex; the value of the bundles assigned is equal to 1, i.e. $p \cdot h(p) = h^{-1}(x) \cdot x = 1$.

4.2. PRIVATE GOODS, PUBLIC GOODS AND EXTERNALITIES

The consumption set X in the previous section was contained in an m-dimensional commodity space. Since only one consumer (or economic agent) was considered, whether the consumption (or use) of a good might affect the well-being (or production) of other agents was not relevant. It is obvious that this aspect is extremely important in an economy with more agents, as it determines the operations which may be applied to commodities. Therefore, this point must be analysed *before* something meaningful can be said about social production and social consumption.

The effect of the consumption (or use) of a commodity by one agent on the well-being (or production) of other agents is called an *external effect* or an *externality* of that commodity. When externalities are present in the economy, it is not sufficient for the commodities to be indexed by their physical properties, they must also be indexed by the economic agent who is consuming (or using) the commodity. Thus, each consumption (or input) variable has three subscripts: x_{kji} stands for the consumption of commodity k by household j as it affects the welfare of houschold i. This model has been already employed by Arrow (1969).

A commodity space in which the goods are characterized both by their physical proporties ($k=1, \ldots, m$) and the consuming (or using) agents ($j=1, \ldots, h$) is called a *personalized commodity space*, R^{hm}. The dimension of the personalized commodity space and its dual space, the *personalized price space*, R^{hm*}, are equal to hm. It will be shown that, under suitable assumptions, the personalized commodity space (with dimension hm) can be reduced to the usual or impersonal commodity space with dimension m (see property 4.2.1). A consumption variable x_{kj} in this space indicates the quantity of commodity k consumed by household j. Let for each consumer $i \in H$, a consumption set $X_i^M \subseteq R^{hm}$ be given, on which a preference ordering $\precsim_i^M$ is defined.

There are *no externalities* in the consumption (or use) of commodity k if each agent is indifferent to the consumption of another agent, i.e. for each $i \in H$, $x_{kii} = y_{kii}$ implies that $(x_{k1i}, \ldots, x_{khi}) \sim_i^M (y_{k1i}, \ldots, y_{khi})$. In figure 4.2.1, the indifference contours of agent i are given for good (ki) and (kj). The price set, P_i^M, and its indifference contours are derived from those in X_i^M by a duality operation (see section 4.1 and the upper polarity operation in section 9.2). The sets X_i^M are scaled such that zero is not an element of these sets.

Fig. 4.2.1. Indifference contours in R^{hm} and R^{hm*} if good k has no externalities in the quantities consumed.

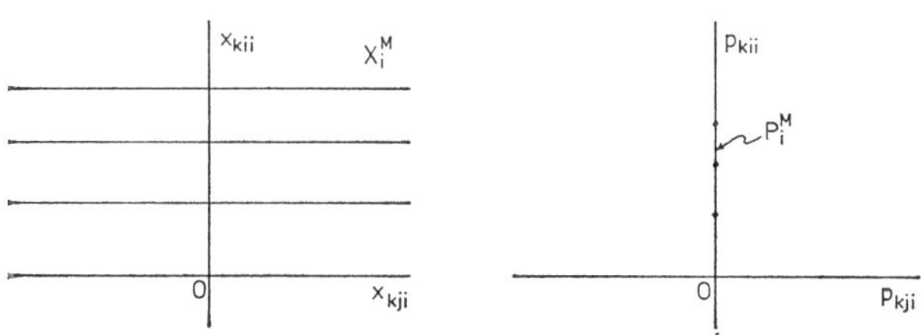

If the consumption (or use) of a commodity k has externalities (see fig. 4.2.2), then these externalities may be personal or impersonal. If agents care not about which particular individuals consume (or use) commodity k, but only about total consumption by others, then commodity k has *impersonal externalities*. Thus, agent i is indifferent for (or has no opinion about) two consumption allocations in which the quantities of consumption are permuted over other agents.

Fig. 4.2.2. Indifference contours if k has externalities

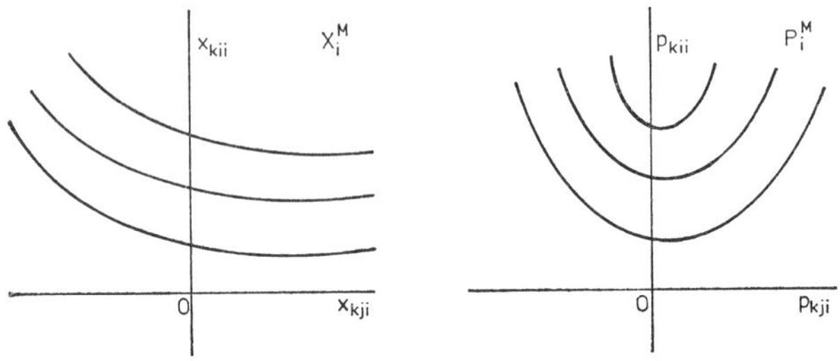

Finally, a commodity k has a *pure externality* if agents have no opinion about the consumption of that commodity by another agent, i.e. $x_{kji} \in X_i^M$ when $j \neq i$ (see fig. 4.2.3). Such a commodity is said to have a pure externality as it leaves the possibility open for the same good to enter into the utility function of more agents (the usual definition of externalities). Therefore, although nobody even realizes that another agent is consuming the same commodity as he is, he may greatly benefit from the consumption of that other agent. A pure externality is by definition impersonal. The price set, P_i^M, and the indifference contours in P_i^M are derived by the upper polarity operation. Since X_i^M is one-dimensional in the subspace of personalized consumption of commodity k, it follows (see section 9.2) that P_i^M is h-dimensional in the subspace of personalized prices of k. When definitions for prices are used analogous to those defined above for quantities, then the following property follows:

Property 4.2.1.

There are no externalities in the consumption (or use) of a commodity k if and only if there are pure externalities in the prices of commodity k.

There are pure externalities in the consumption (or use) of a commodity k if and only if there are no externalities in the prices of commodity k.

If no goods have any externalities in an economy, then every agent i is indifferent between the bundle $(x_{k1i}, \ldots, x_{kii}, \ldots, x_{khi})$ and the bundle $(0, \ldots, x_{kii}, \ldots, 0)$ in which consumption by other agents is put equal to zero. Therefore, from the point of view of consumer i, these bundles are sufficiently identified by the variable $x_{kii,}$ which can be reduced to x_{ki}. In this case, the dimension hm of the i-th consumption set X_i^M in the hm-dimensional personalized commodity space can be reduced to m. If all goods have pure externalities in an economy, then $x_{kji} \notin X_i^M$ and the dimension of X_i^M is equal to m. Thus, the following property is established:

Property 4.2.2.

If all goods in an economy have either pure or no externalities, then the personalized commodity space with dimension hm can be reduced to the impersonal commodity space with dimension m.

The relation between X_i in R^m and X_i^M in R^{hm} is expressed by the following definition:

$$X_i := \text{Proj}_i \, X_i^M, \quad \text{where}$$

the i-th projection of X_i^M is done in the commodity space R_i^m, in which the quantities consumed by i are indicated. The effects of the i-th consumption on j can thus never be expressed in R^m (or X_j), but only by X_i^M in R^{hm}.

Fig. 4.2.3. Indifference contours for a good k with pure externalities in the quantities consumed

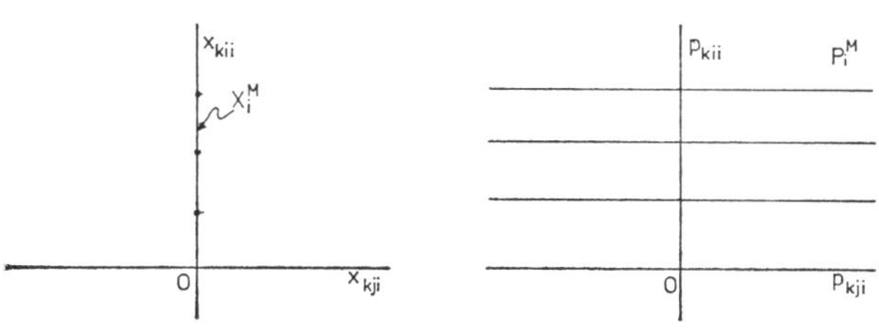

Next, those characteristics of a commodity are defined which are based upon its technical or 'objective' properties of a commodity. A commodity is said to be *exclusive* if consumption (or use) by one agent excludes consumption by any other agent. In this case, total or social consumption is equal to the sum of individual consumption: $x_k = \sum_j x_{kji}$; and a specific quantity can be consumed either by one or by another agent. A commodity is said to be *collective* in relation to an economy if all agents in the economy consume (or use) that commodity. In this case, total or social consumption is equal to the consumption of each individual: $x_k = x_{kj}$;

and a specific quantity of that commodity can be consumed by one and by every other agent in the economy.

The 'subjective' properties of a commodity attributed to it by the agents and the 'objective' technical properties are combined to define the following polar cases:

A commodity is said to be a *private good* in an economy if it is an exclusive good without externalities.

A commodity is said to be a *public good* in an economy if it is a collective good with pure externalities relative to the economy.

It follows from property 4.2.2. that an economy with private and/or public goods can be defined in the impersonal commodity space.

4.3. PRIVATE GOODS, PUBLIC GOODS AND TRANSACTION COSTS

Two kinds of definitions of a private good need to be distinguished: firstly, model-oriented definition stating the conditions which permit application of a market-mechanism (the definition given in section 4.2); secondly, the institution-oriented definition stating as a condition that a market-mechanism be used in the allocation of a commodity.

Since the model-oriented definition is very narrow (most goods do have some externalities in consumption), the market can be applied either to a small set of economic goods giving an optimal allocation, or to a larger set of goods (which are not all strictly private goods) giving approximately an optimal solution. The latter choice is made in economic life, as the costs of allocation-mechanisms are not zero. If the commodities deviate too much from the conditions set for private goods, then characteristics are added (at some cost) to allow for treatment by a market, such as the exclusion of other consumers by means of fences, laws or morals. These costs belong to the category of transaction costs, to be defined below.

Therefore, the institution-oriented definition is close (or equal) to the common sense concept of a private good, but if one wants to analyze the economic system and its allocation-mechanisms, one must use the model-oriented definition.

An analogous situation can be found in the theory of public goods. For example, an institution-oriented definition of public goods has been given by Buchanan (1968, p. 1): 'public goods are goods that people are observed to demand and supply through political institutions'. The pragmatism of this definition is attractive as it suggests that if the sufficient conditions for action through political mechanisms were not fulfilled, other mechanisms would be found. Definitions along the same line as those of Buchanan's have been given by Musgrave (1969, p. 125) and Barone (1912, p. 165). However, when attempting to analyze which mechanism is best suited for the allocation of specific commodities in an economy, the model-oriented definition is to be preferred.

The definition of a public good given above (section 4.2) is narrower than that given by Samuelson (1954, p. 387): '... collective consumption goods $(x_{n+1}, \ldots, x_{n+m})$ which all enjoy in common in the sense that each individual's consumption of such a good leads to no substraction from any other individual's consumption of that good, so that $x_{n+j} = x^{i}_{n+j}$ simultaneously for each and every i-th individual and each collective good'. This definition allows for positive external effects of some individual's consumption on another's consumption, which is excluded in section 4.2. The definition used here is, however, narrower than that later given by Samuelson (1966, p. 102) when he says: 'a public good is simply one with the property of involving a 'consumption externality' in the sense of entering into two or more persons' preference functions simultaneously'. This definition corresponds with a commodity having an external effect in consumption, as defined in section 4.2.

The definition of public goods given here implies a number of characteristics on which public goods have been defined elsewhere, or which follow from other definitions. To be more specific: 'indivisibility' of public goods is essential for Sax (1887), Bowen (1943, p. 27) and Drees e.a. (1968, p. 15); 'non-excludability of use (or consumption)' is an essential characteristic for Dorfman (1966, p. 248), Stevers (1967, p. 41) and Milleron (1972, p. 424). Both aspects are required by Johansen (1965, p. 17), while Foley (1967, p. 49) proposes 'involuntary consumption' as essential. These characteristics follow from the condition required here that a public good be collective, i.e. be consumed or used by every agent in the economy.

Head (1962, p. 205) and Musgrave (1969, p. 126) add to this type of characteristic another one which is equivalent to pure externalities; they call it 'external economies to an extreme degree', resp. 'non-rivalness in consumption'. This point of view lies close to Samuelson's earlier definition.

A different approach is suggested by Arrow (1970). He states that the concept of public goods belongs to the more general one of 'externality', which in its turn belongs to the broader category of 'market failure'. Market failure, again, is not absolute: 'it is better to consider a broader category, that of transaction costs, which in general impede and in particular cases completely block the formation of markets. It is usually though not always emphasized that transaction costs are costs of running the economic system'. Transaction costs relative to the market-mechanism are, for example: the costs of implementing the exclusivity of consumption of private goods; the costs of information needed to enter and participate in any market; the costs of preventing externalities in consumption, such as disposal facilities.

In general, I propose to define the *transaction costs* relative to some allocation mechanism as being the costs of adding characteristics to a commodity necessary to apply that allocation mechanism within the economic system. The transaction costs relative to a market of buying a refrigerator, for example, include time and money spent on information about alternatives, costs related with the exchange-commodity money, costs of ultimate disposal after consumption. It is not very hard to find examples in which transaction costs surpass the price of the commodity itself. Imagine the transaction costs relative to a market of an hour's broadcasting; or of police services in city-traffic. In such cases, it seems plausible to look for another allocation mechanism.

On the other hand, transaction costs relative to a political mechanism can also be prohibitively high for many commodities. It was Hayek (1935) who draw attention to the tremendous communication and uncertainty problems related to the socialist organization. The costs of a bureaucracy will for a large part consist of the transaction costs relative to a political mechanism, and even then the real transaction costs might be higher.

Tinbergen (1961) contends that the optimal organization of economic decisions is one of the main features in modern welfare theory. This statement can be made more specific by putting the following questions:

1. how many and which allocation-mechanisms should operate within an economic system?
2. which commodities are allocated by which allocation-mechanisms?

The first question is relevant as the costs of any allocation-mechanism are not zero. Although under certain conditions and at certain costs, goods with external effects can be transformed into private goods or public goods, the production of many goods in an economic system depends on the availability of a suitable allocation-mechanism. The answer to the first question, therefore, directly influences the choice of individual consumers and their welfare; but the answer is hard to give, as one must compare the transaction- and disutility-costs of producing one bundle of goods with the transaction costs and the costs of the allocation mechanism in producing another bundle of goods. It is still the only way to compare and valuate various economic systems for a given economy and its members.

The second question can be solved by comparing the transaction costs of a commodity relative to the allocation-mechanisms in the system. But since these costs are hard to make explicit, it is not easy to judge whether a choice is socially efficient. However, before a decision is made to supply some good through the market-mechanism and the political-mechanisms simultaneously, it is necessary first to analyze the over-all effects.

Although the various allocation-mechanisms ought to be considered as being themselves public goods, the hypothesis to be made that all goods are either private or public makes it possible to ignore the costs of allocation-mechanisms and the transaction costs. The restrictions stemming from the hypothesis are thus rather strong.

4.4. SOCIAL CONSUMPTION AND SOCIAL PRODUCTION

The definitions of private and public goods given in section 4.2 imply that the personalized commodity space can be reduced to the impersonal commodity space only when these goods are present in the economy (see property 4.2.2).

Let R^m be the subspace of private goods and let R^n be the subspace of public goods. Let consumer i's consumption bundle be $(x_i, z_i) \in R^{m+n}$ for $i \in H$, the set of consumers in the economy. Then *social consumption* in the economy is equal to $(x, z) \in R^{m+n}$, where $z = z_i$, for all i, and $x = \Sigma x_i$. The *social consumption set* $X \subseteq R^{m+n}$ is equal to

$$\{ \sum_{i \in H} (\text{Proj}_m X_i), \bigcap_{i \in H} (\text{Proj}_n X_i) \}.$$

Since the utility functions are usually assumed to be index functions which are not comparable between consumers (ordinal utility), it is not possible to aggregate these functions to obtain social utility or social welfare. However, aggregation of the upper-preference sets over individuals is possible, if only the allocation that determines the upper-preference sets and the weight of each consumer in aggregation are indicated (see property 5.4.2). In this way, a necessary condition for maximal social welfare can be obtained.

The situation is different in production. The input and output of a production process are assumed to permit exact measurement and different processes can therefore be compared and aggregated. Further, it must be stressed that criteria for the private or public character of commodities are to be found at the input side of a consumption or production process. It is the consuming or using agent who determines whether there exist external effects in the consumption or use of commodities by other agents. Whatever the character of the output of processes as inputs in the following process, it will be assumed that these outputs are additive over the various processes. When the output of two production units is, for example, roads and smoke, then the output of both units together is equal to the sum of each (road and smoke) output, although both commodities are probably public goods.

Two cases will be distinguished: production in an economy having only private goods, and the more general case of production in an economy having also (or only) public goods. In an economy with private goods only, *the social production set* Y is equal to the sum of the production sets Y_j of the various units j. It should be noticed, however, that this definition does not imply that the input-output multifunctions (see section 3.1) of the units can be added to obtain the social input-output multifunction $f(x) := \{y \mid (-x, y) \in Y\}$. For, assume that

$$f_1(x) := \{y \mid (-x, y) \in Y_1) \quad \text{and}$$
$$f_2(x) := \{y \mid (-x, y) \in Y_2)$$

then the sum $f_1 + f_2$ is defined by (see section 10.1)

$$(f_1 + f_2)(x) := f_1(x) + f_2(x)$$
$$= \{y_1 + y_2 \mid (-x, y_1) \in Y_1 \quad \text{and} \quad (-x, y_2) \in Y_2\}.$$

It follows immediately that the graphs are not equal:

$$G(f_1 + f_2) \neq G(f_1) + G(f_2) = Y_1 + Y_2 = Y = G(f).$$

The appropriate operation to perform on the individual input-output multifunctions to obtain the social input-output multifunction is the operation of disjunction, $\vee$, defined in section 10.1:

$$(f_1 \vee f_2)(x) := \{f_1(x_1) + f_2(x_2) \mid \exists x_1, x_2 : x = x_1 + x_2\}.$$

Since $G(f_1 \vee f_2) = G(f_1) + G(f_2) = Y_1 + Y_2 = Y = G(f)$, it follows that $f = f_1 \vee f_2$, i.e. the *social input-output multifunction* is equal to the disjunction of the individual input-output processes: $f := \vee_{j \in F} f_j$. (For the properties of disjunction under the inverse or duality operations, see chapter 10.)

The other approach is based on partition of the commodity space into private and public goods, valid for all participants in the economy. This includes inputs, outputs and intermediate goods in an economy with private and/or public goods. Since the input of public goods in a production process is equal for all producers in an economy but the input of private goods will vary, a kind of production multifunction will be defined which associates for each producer the net output of private goods and the total output of public goods with the total input of public goods in the producer's production technology. The convention regarding to signs is (as in chapter 3) that inputs have negative signs and outputs positive signs.

Therefore, let $(x_j, z_j) := (x_{kj}, z_{lj})$ be a vector in the commodity space R^{m+n}, indicating the quantities of private goods ($k \in M$) and public goods ($l \in N$) for producer $j \in F$. The multifunction Y_j: $R^n \rightarrow R^{m+n}$ is called the j-th producer's *production-multifunction* if for each public good input $z \in R^n$, the image set $Y_j(z)$ indicates the production possibilities for producer j. In fact, the producer can choose any point in $Y_j(z)$, and so determine what quantities of private goods he buys and what quantities of private and public goods he sells. The input of public goods, z, is given for each producer. It is evident that if all goods in the economy are private goods, the dimension of R is zero and $Y_j(0) = Y_j$, the production set defined in section 3.1.

In an economy with both private and public goods the *social production set* $Y(z)$, given the bundle of public goods, is equal to the sum of the individual production sets $Y_j(z)$; i.e. $Y(z) := \sum_{j \in F} Y_j(z)$. The composition of the components of a social product vector and the allocation of the private goods can not yet be determined. Only the technically feasible alternatives in the composition of social output are given by $Y(z)$.

The corresponding production cost-prices for each producer with production set $Y_j(z)$ are determined by the polarity operation on $Y_j(z)$, as is done in section 3.3. The j-th producer's *production-price-multifunction* $^*Y_j : R^n \rightarrow R^{m+n*}$ is defined by

$$^*Y_j(z) := [Y_j(z)]^*_-, \quad \text{for each} \quad z \in R^n.$$

This indicates the composition of cost-prices relative to producer j-th's technology when the input of public goods z is given, so that the value of output (producer's turnover) is at the most equal to 1. These prices are still personalized prices as they are uniquely determined by the j-th producer's technology.

The *social production-price set* $^*Y(z)$ is determined by the multifunction $^*Y: R^n \to R^{m+n*}$ defined by:

$$^*Y(z) := [Y(z)]^*_- = [\sum_{j\in F} Y_j(z)]^*_-, \quad \text{for each} \quad z.$$

This multifunction is the polar of the social production-set multifunction $Y(z)$ (see section 10.1) and determines the social cost-prices of each commodity in the economy when the input of public goods z is given, so that the value of output (net national product) is at the most equal to 1. These prices are impersonal prices and are scalar multiples of the personalized prices in $^*Y_j(z)$; the values of the scalars λ_j determine the income distribution ovei the producers, or better, the contribution of each producer to the net national product. This follows from the following property (based on property 9.3.4.1):

Property 4.4.1.

If for each $z\in R^n$ the production sets $Y_j(z)$ are closed, convex and contain zero, then

$$^*Y(z) = [\sum_{j\in F} Y_j(z)]^*_- = \overset{\circ}{\bigcap_{j\in F}} [Y_j(z)]^*_- = \overset{\circ}{\bigcap_{j\in F}} {}^*Y_j(z).$$

The $\{\lambda_j\}$ follow from the definition of convex intersection (see definitions in sections 8.3 and 9.3):

$$\overset{\circ}{\bigcap_{j\in F}} {}^*Y_j(z) := \left\{ q \,\middle|\, \begin{array}{l} \exists q_j \in {}^*Y_j(z),\ \exists\lambda_j\in[0,1]: \\ \sum\lambda_j = 1 \text{ and } q = \lambda_j q_j,\ \forall j\in F \end{array} \right\}.$$

Until now, the input of public goods z has been accepted as given by the individual producers. But each producer has a definite opinion about what is the 'best' composition of the public goods vector z, which co-determines his set of production possibilities $Y_j(z)$. This opinion is expressed by the adjoint multifunction of the producer's production multifunction, $Y_j^*: R^{m+n*} \to R^{n*}$, defined below, and is based on the prices of his decision parameters: the net output of private goods and total output of public goods.

The j-th producer's (public-goods) *benefit multifunction* $Y_j^*: R^{m+n*} \to R^{n*}$ is the adjoint of his production multifunction Y_j, defined by (see also section 10.1):

$$Y_j^*(p,q) := \{r_j | \forall z\in R^n, \forall (x,y)\in Y_j(z): r_j z \geqq px+qy-1\}.$$

This indicates in which direction expansion (or, rather, change) of the various components of the public goods input vector is most profitable for the j-th producer.

The public-goods *producer's benefit multifunction* $Y^* : R^{m+n*} \to R^{n*}$ is the adjoint of the social production multifunction $Y = \sum_{j \in F} Y_f$. It gives the weighted average of the producer's valuations r_j, the weights being equal to each producer's contribution to the net national product when maximized at prices (p, q). This follows from the following property (derived from property 10.4.6.1):

Property 4.4.2.

If the producer's production multifunctions Y_j are convex cone-interior processes having the same orientation (defined in section 10.4), then

$$Y^* := [\sum_{j \in F} Y_j]^*_+ \supseteq \overset{\circ}{\underset{j \in F}{+}} Y_j^*,$$

$$\text{where } \overset{\circ}{\underset{j \in F}{+}} Y_j^*(p, q) := \bigcup_{\lambda_j} \left\{ \sum \lambda_j \, Y_j^*(p_j, q_j) \,\middle|\, \begin{array}{l} \lambda_j \in [0, 1] : \sum \lambda_j = 1, \text{ and} \\ (p, q) = \lambda_j (p_j, q_j), \quad \forall j \in F \end{array} \right\}.$$

Thus, with each social cost-price system (p, q) of private goods and public goods can be associated a set of vectors of valuations of public goods inputs, as valuated by the production side of the economy.

A similar valuation of the public goods input should also be given for the consumption side of the economy. Its formulation, however, requires aggregation of consumers' upper-preference sets which cannot be done without defining equilibrium in the economy.

4.5. AN ECONOMY WITH LOCAL PUBLIC GOODS

In the previous section it was assumed that a public good x_k can be consumed or used by each agent in the economy: $x_{ki} = x_k$ for each $i \in H \cup F$. This assumption can be weakened by introducing consumption collectivities in the economy for each good, which determine the extent of public use for each good. A subset of agents in the economy is called a *collectivity*. An economic good k is said to be a *local public good* relative to a collectivity $G_{kg} \subseteq H \cup F$, if it is a private good in the economy for all agents except those within the collectivity, for whom it is a public good.

The collectivity H_{kh} is called the h-th *consumption-collectivity* with respect to local good k; the *index set of consumption-collectivities* with respect to local good k is indicated by $H_k = \{l, \ldots, h, \ldots, \bar{h}_k\}$, and the

h-th consumption-collectivity of which agent i is a member is indicated by H_{khi} (see fig. 4.5.1).

It follows from the definitions that for each local good k, the set of consumers H is partitioned in consumption-collectivities $\{H_{kh}\}$:

$$\bigcup_{h \in H_k} H_{kh} = H \qquad \text{and} \qquad \bigcap_{h \in H_k} H_{kh} = \phi .$$

The number of consumption-collectivities for good k, $\bar{h}_k$, is equal to the cardinal number of the set H_k; this number is equal to the number of *consumption-units* of local good k supplied by the producers $\{x_{kh}\}$.

Examples of consumption-collectivities (with examples of their local goods) are: households (furniture); parishes (church); suburbs (shopping centers); cities (fire protection); sport clubs (play-grounds); theater visitors (movie); subscribers (information); the United Nations (international law and order); etc.

For each member i of a consumption-collectivity H_{kh}, consumption is given by: $x_{ki} = x_{kh}$.

Fig. 4.5.1. Consumption-collectivities of local public good k

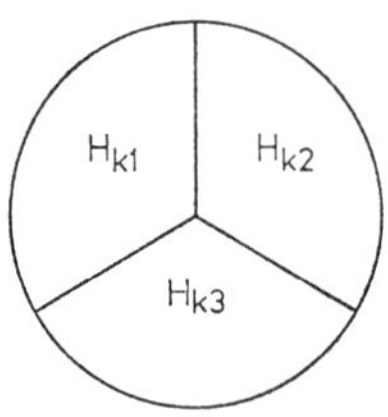

$H = H_{k1} \cup H_{k2} \cup H_{k3}$
$H_k = \{1, 2, 3\}$
$\bar{h}_k = 3$

In an economy with local public goods, *social consumption* is defined as being equal to the summation of consumption by collectivities: $x_k := \sum_{k \in H_k} x_{kh}$, for every k.

From the definitions given above, it follows that private goods and public goods are extreme cases of local public goods. Let k be a private good, then:

— each consumer is equal to a consumption-collectivity, i.e. $H_{khi} = H_{ki} = \{i\}$, for each i;
— the set of consumption-collectivities H_k equals $H = \{1, \ldots, \bar{h}\}$;
— the number of consumption-units $\bar{h}_k$ equals $\bar{h}$;
— social consumption equals $x_k = \sum_{i \in H} x_{ki}$.

Let k be a public good, then:

— the set of consumers is equal to the consumption-collectivity, i.e. $H_{khi} = H_{ki} = H$;
— the set of consumption-collectivities H_k equals $\{1\}$ and contains one element, i.e. $\bar{h}_k = 1$;
— social consumption equals $x_k = x_{ki}$, for each $i \in H$.

The *social consumption set* X of the economy is obtained through the following aggregation of individual consumption sets X_i:

$$X := \left\{ (x_1, \ldots, x_k, \ldots, x_m) \middle| \begin{array}{l} \exists (x_{1i}, \ldots, x_{ki}, \ldots, x_{mi}) \in X_i, \\ \qquad\qquad\qquad \exists H_{khi} = H_{kh} \subseteq H: \\ x_{kh} = x_{ki} \quad \text{and} \quad x_k = \sum_{k \in H_k} x_{kh}. \end{array} \right\}$$

It can be easily seen that these definitions comply with those given in section 4.4.

Analogous definitions hold for the inputs of local goods in a production process.

The collectivity F_{kf} is called the f-th *production-collectivity* for local good k, if k is a private good in the economy for all agents except for the producer-members of the collectivity, for whom input k is a public good. The *index-set of production-collectivities* is indicated by $F_k = \{1, \ldots, f, \ldots, \bar{f}_k\}$, and the f-th production-collectivity for input k of which producer j is a member is indicated by F_{kfj}.

Again, the set of producers F is, for each local public input good k, partitioned in production-collectivities $\{F_{kf}\}$. Their number, $\bar{f}_k$, is equal to the number of *production-units* of good k, which is an intermediate product by definition.

Examples of production-collectivities (with examples of an associated local public goods) are:
joint insurance agreements (covering disasters);
shopkeepers' combinations (attraction of buyers);
unions (indivisibility of labor);
associations of registered professions (quality of product);
corporations (economies of scale);
integrated concerns (control of raw and intermediate material);
agglomerates (profit stabilization);
hotel chains (reservations).

It should be emphasized, however, that none of the examples perfectly meets the requirements of the definition of a local public good. Even if the 'objective criteria' of private goods (exclusivity) and public goods (collectivity) are met, then the 'subjective criteria' are hard to accomplish.

The input of local public goods determines the set of production possibilities for each producer. To this extent, the definition of the production multifunction Y_j given in section 4.4 can also be applied here. However, since local goods do not permit addition of the individual multifunctions to obtain the social production multifunction, it is preferable to use the input-output multifunction $f_j : R^n \to R^n$ to describe the j-th producer's technology.

For the case of an economy with local public goods, the *social input-output multifunction* $f : R^n \to R^n$ is defined by:

$$f(x) := \left\{ \sum_{j \in F} f_j(x_{1j}, \ldots, x_{kj}, \ldots, x_{nj}) \;\middle|\; \begin{array}{l} \exists F_{kfj} = F_{kf} : x_{kj} = x_{kf} \quad \text{and} \\ x_k = \sum\limits_{f \in F_k} x_{kf} \end{array} \right\}$$

The *social production set* Y is then defined by

$$Y := \{y - x \mid y \in f(x)\}.$$

Finally, an economy with local public goods ε is given by:

$$\varepsilon := \{H, (X_i, \precsim_i, \lambda_i, H_{khi}); F, (f_j, F_{kfj}); w \in R^n\}.$$

5. Equilibrium in economies with private and public goods

5.1. THE VALUATION REPRESENTATION OF AN ECONOMY

The preferences of consumers as well as the technology of producers are usually expressed as relations on the commodity space in which quantities are given for each commodity (see sections 3.1 and 4.1). However, if a common constraint is given, such as the value in which a bundle of quantities has to be mapped, then the relations characterizing consumers and producers can also be expressed in terms of the linear mappings, such as the prices. This (mathematical) operation is called a duality operation (see section 9.2). In this way, information about the economy in terms of quantities is carried over into information in terms of prices or valuations. If it is possible, through the duality operation, to restore the original information in terms of quantities, then no information is lost during the transformation processes. In this case, the economy represented in terms of quantities can be considered as a representation in terms of prices, and vice-versa. This means that one can express properties of individual agents or of the economy either in quantities or in valuations, which ever best suits the set purpose, and translate these concepts into the other space.

Since in the case of private goods the conflict of interest between agents is concentrated on quantities, the most natural definition of equilibrium will be given in the quantity space (although translation in terms of prices is possible). On the other hand, in the case of public goods the conflict of interest between agents concentrates on the valuations; in this case it is therefore natural to define equilibrium in the price space.

Let the distribution economy (introduced in section 2.2)

$$\varepsilon := \{H, (X_i, \precsim_i, \lambda_i); F, (Y_j); w \in R^{m+n}\},$$

be defined by a set H of consumers, each having a preference relation on his consumption set, and some weight λ_i; a set F of producers, each having a typical production multifunction Y_j; and a vector of initial resources w partitioned in private goods: $w_M \in R^m$, and public goods: $w_N \in R^n$.

Let the economy in the price space,

$$\varepsilon^* := \{H, (P_i, \lesssim_i^*, \lambda_i); F, (^*Y_j); (w)^* \subseteq R^{m+n*}\},$$

be defined by a set of consumers, each having a costing relation on his set of consumption-prices P_i (see section 4.1), and some weight λ_i; a set F of producers each having a production technology expressed by cost-price multifunction $^*Y_j$; and a hyperplane $(w)^*$ indicating the prices corresponding with the initial resources of private and public goods, and corresponding with the vector w in ε.

The economy ε^* is said to be a *polar economy* of ε if the sets and relations in R^{m+n*} are derived from sets and relations in R^{m+n} by means of the polarity operation *, defined in sections 9.2 and 10.1 (see section 4.1 and 4.4).

The polar economy ε^* is said to be a *valuation representation* of the economy ε if ε is also a polar economy of ε^*. The following property follows directly from the definitions of P_i, $\lesssim_i^*$ and $^*Y_j$ given in sections 4.1. and 4.4, and from the properties of reflexive sets given in section 9.3:

Property 5.1.1.

Let ε^* be the polar economy of the economy ε defined above. ε^* is a valuation representation of ε if and only if:

1. The upper-preference sets $C_i(x)$ are aureole-reflexive for each i and for each $x \in X_i$, i.e. they are closed, convex, aureoled and do not contain 0;
2. The production sets $Y_j(z)$ are star-reflexive for each j and for each $z \in \mathrm{Dom}\,(Y_j)$, i.e. they are closed, convex and contain 0.

Important properties of the concepts in ε and ε^* are invariant under the polarity operation *. Consider the following conditions on concepts in ε^*.

Conditions 5.1. For each consumer i∈H:

5.1.1. The price set $P_i = \mathrm{Dom}\,(\mathrm{D}_i)$ is convex.
5.1.2. The price set P_i is completely preordered by $\lesssim_i^*$
5.1.3. The costing relation $\lesssim_i^*$ is convex.
5.1.4. The indifference sets are thin.
5.1.5. All choice sets in P_i and generated by elements of P_i have maximal elements.

5.1.6. The costing relation $\precsim_i^*$ is monotone and $P_i \subseteq R_+^{m*}$.
The following property can be derived from property 4.1.3:

Property 5.1.2.

Consider the distribution economy ε and its polar economy ε^*. The assumptions 4.1 are satisfied for each consumer in ε if, and only if, the conditions 5.1 hold in ε^*.

This property also holds if assumption 4.1.6 and condition 5.1.6 are simultaneously omitted. Similar properties can also be derived for production: see properties 5.1.1.2, 9.3.1 and 10.2.5. If a producer's multifunction Y_j is monotonously increasing, then his cost-price multifunction $^*Y_j$ is nonnegative. It is evident that if all production sets are star-reflexive, and assumptions 4.1 (or conditions 5.1) are satisfied, then the polar economy ε^* is a valuation representation of ε.

One of the virtues of a valuation representation of an economy is that one can prove properties of the economy in the quantity space or in the price-space, whichever is most appropriate in the given circumstances. This property of a valuation representation has been used, for example, by Weddepohl (1972, 1973a) to formulate a rather simple proof of the existence of an equilibrium for a private goods economy in the price space.

In order to compare definitions of the same concepts in the quantity space and in the price space, the definitions given in chapter 2 for a private goods economy $E_1 \in \varepsilon$ are translated for its polar economy $E_1^* \in \varepsilon^*$.

Let $E_1 := \{H, (X_i, \precsim_i, \lambda_i);\ Y;\ R^n\}$ be a distribution economy for private goods only, with a set of H consumers, each having a consumption structure following the assumptions under 4.1 and a positive weight λ_i in the income distribution; the production is given by a social production set which includes the vector of initial resources and which is closed and convex, and which also contains zero and the non-positive orthant. According to property 5.1.1, this economy has a valuation representation:

$$E_1^* = \{H, (P_i, \precsim_i^*, \lambda_i);\ Q;\ R^{n*}\}.$$

For this economy, the following property can be derived from properties 5.1.1, 5.1.2 and 9.4.3.2:

Property 5.1.3.

Let the economy E_1 have a valuation representation E_1^* and let the recession cone of the social production set, $Y(w)$, given w, be such that $0^+ Y(w) \cap R_n^+ = \{0\}$ and $\text{Int}\ Y(w) \cap R_n^+ \neq \phi$, then the following four concepts are defined equivalently by (E_1) and by (E_1^*):
An allocation (x_i, y) in E_1 is an attainable state or a *feasible allocation* if and only if one of the following conditions hold in E_1, resp. E_1^*: (see also fig. 5.1.1)

$$(E_1)\ x_i \in X_i \quad \text{and} \quad \sum x_i = y \in Y;$$

$$(E_1^*)\ M_i(x_i) \subseteq P_i, \quad \text{and} \quad \mathring{\cap}\ M_i(x_i) = M(y) \supseteq Y^*.$$

An allocation (x_i, y) is an *efficient allocation* if and only if

(E_1) $x_i \in X_i$ and $\sum x_i = y \in \text{Bnd}\ Y$;
(E_1^*) $M_i(x_i) \subseteq P_i$ and
$H(y)$ supportingly separates $\mathring{\cap}\ M_i\ (x_i)$ and Y^*.

Therefore, if (x_i, y) is an efficient allocation, then there exists an arbitrary income distribution $\{\tilde{\lambda}_i\}$, and prices $\{p_i, p\}$ such that

(E_1) $x_i \in M(p_i)$ and $\sum x_i \in M(p) \cap \text{Bnd}\ Y$;
(E_1^*) $p_i \in H(x_i)$ and $\tilde{\lambda}_i p_i = p \in H(y) \cap \text{Bnd}\ Y^*$;

where $\tilde{\lambda}_i$ is an income distribution, i.e. $\tilde{\lambda}_i > 0$ and $\sum \tilde{\lambda}_i = 1$.

An allocation (x_i, y) is a *Pareto optimum* in E_1 if and only if it is an efficient allocation and there exists a price p with income distribution $\tilde{\lambda}_i$ such that:
(E_1) the choice sets $M_i(p;\ \tilde{\lambda}_i)$ support $C_i(x_i)$ in x_i;
(E_1^*) the choice sets $M_i(x_i)$ support $^*C_i\ (x_i)$ in $p_i = p/\tilde{\lambda}_i$.

The income distribution required for a Pareto optimum need not to be equal to the one which is given in E_1, resp. E_1^*. If both are equal, however, then the Pareto optimum is a competitive equilibrium, which permits decentralization of consumption and production decisions on the basis of a price vector:

A state vector (x_i, y, p) in E_1 is said to be a *competitive equilibrium* in E_1 if and only if either:
(E_1) $H(p)$ supportingly separates $\Sigma C_i(x_i)$ and Y in y; or:
(E_1^*) $H(y)$ supportingly separates $\mathring{\cap}\ ^*C_i(x_i)$ and Y^* in p.

Fig. 5.1.1. A competitive equilibrium in the commodity and price spaces

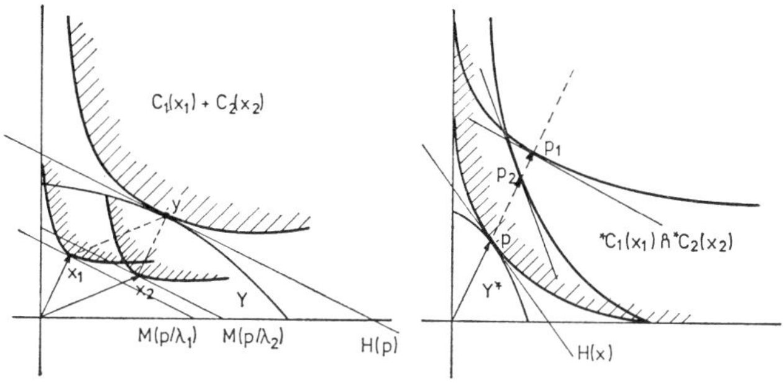

This condition is equivalent to, resp.:

(E_1) a. $M_i(p; \lambda_i)$ supports $C_i(x_i)$ in x_i, for each i;
b. $H(p)$ supports Y in y;
c. $y = \Sigma x_i$; or:

(E_1^*) a. $M(x_i)$ supports $^*C_i(x_i)$ in p_i, for each i;
b. $H(y)$ supports Y^* in p;
c. $p = \lambda_i p_i$, for each i.

It follows that a competitive equilibrium results in an allocation in which (a) all consumers are in equilibrium, (b) the producers are in equilibrium and (*c*) the market is in equilibrium. Each of these conditions can be expressed by means of choice sets in the quantity space, such as $M(p; \lambda_i)$, or choice sets in the price space, such as $M(x_i)$. The second alternative, however, has hardly any economic significance; this is in contrast to the case of public goods.

5.2. EQUILIBRIUM IN AN ECONOMY WITH PUBLIC GOODS ONLY

The first definition of equilibrium in an economy with public goods was presented by Lindahl (1919). His idea was to extend the competitive equilibrium by introducing personalized prices, the sum of which prices was to equal the production price. These personalized prices were to play the same role as market prices for private goods and thus generate 'demand' multifunctions for public goods. Equilibrium was then established if the demand for a public good by consumers was equal for all consumers, at individual prices the sum of which was equal to the production price (see fig. 5.2.1).

Fig. 5.2.1. Equilibrium 'demand' for a public good by two consumers

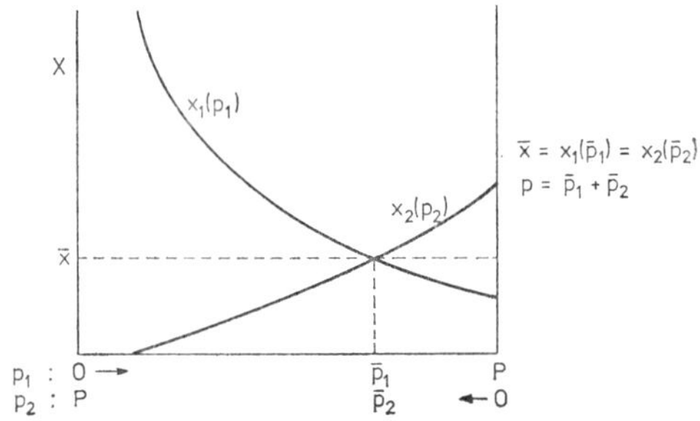

Lindahl's equilibrium concept (defined below) is thus an extension of the competitive equilibrium defined in section 5.1. Mathematically, this extension follows naturally; from the economic point of view, however, many problems arise as these personalized prices are supposed to be given for each consumer, or should be revealed by the inverse demand multi-function.

The rationality of consumers, according to which they are supposed to reveal their individual demand for private goods, i.e. the best and cheapest for each, prevents the consumer spontaneously revealing the price he should, in reality, pay for a public good supplied in the economy.

Therefore, another concept of equilibrium is defined here and called a public equilibrium. The idea is that the private and public sectors can be separated, each having in principle its own prices and its own concept of rationality in behavior. The income distribution for private goods is usually expressed by differences in the individual choice sets, and the 'income' distribution for public goods can be expressed by the weights given to each personalized price, those prices being chosen from a choice set which is equal for all agents. This means, for example, that personalized prices need not necessarily be considered as taxes through which public goods are financed, but may also be considered as a kind of vote – though a complex one (see table 1.4.1).

An economy will thus be partitioned here in terms of consumers' decisions about the allocation of commodities. Two standards of value, two income-distributions and two allocation mechanisms will be introduced, for each level one.

On the level of private goods, a competitive equilibrium is determined via a market-mechanism, given the income distribution $\{\lambda_i^M\}$ in terms of money. On the level of public goods, a public equilibrium is determined via a referendum-mechanism, given the weight distribution $\{\lambda_i^N\}$ in terms of votes.

This partition of an economy into a level of private goods and a level of public goods is inconsistent with the concept of a Lindahl equilibrium, precisely because this concept is an extension of a competitive equilibrium. Thus, the allocation of public goods should be determined via the market-mechanism, given an income distribution $\{\lambda_i\}$ for both private and public goods in terms of money. Since this solution contradicts the assumption of individual rationality, other mechanisms have been proposed (see section 6.4) which have – presumably – better performance characteristics and also arrive at a Lindahl equilibrium, although the income distribution changes during the process to allow for compensatory payments.

The allocation, however, with its changing income distribution, should be called a Pareto optimal allocation, rather than a Lindahl equilibrium. It follows from the dual conditions for a Pareto optimum in a convex economy with public goods (see property 5.2.1.2), that with each Pareto optimum a Lindahl equilibrium can be associated if the income distribution is adapted. Therefore, the set of Lindahl equilibria generated by all possible income distributions is equal to the set of Pareto optima. Since the concept of a Lindahl equilibrium is connected with an initial distribution of resources, or with a given income distribution, the conclusion follows that no satisfactory allocation mechanism has been designed via which a Lindahl equilibrium can be obtained.

It should also be noted that Foley (1970) has shown that the Lindhal equilibrium is an element of the core of an economy. This is true if the external effects are impersonal and fall under some restriction of individual property rights, as has been indicated by Starrett (1973).

In this section, the public equilibrium for an economy with public-goods-only will be defined and its existence will be proven. Formally, a public equilibrium is equivalent to a Lindahl equilibrium, but for reasons indicated above, another name is preferred. Another reason is that people can only be compensated individually when at least one private good is present in the economy. The situation in an economy with public-goods-only is therefore principally different from the situation in an economy with public and private goods.

Consider an economy E_2 with h consumers, each characterized by a consumption structure, f producers, each having a production multi-function Y_j, and n public goods with initial resources $w \in R^n$:

$$E_2 := \{H, (X_i, \precsim_i, \lambda_i); F, (Y_j); w \in R^n\} .$$

The polar economy is then:

$$E_2^* = \{H, (P_i, \lesssim_i^*, \lambda_i); F, (^*Y_j); \{w\}^* \subseteq R^{n*}\}.$$

A state vector (x, p_i^L, p) is called a *Lindahl equilibrium* for E_2 if:
(a) the choice sets $M(p_i^L; \lambda_i)$ support $C_i(x)$ in x, for each i;
(b) the hyperplane $H(p)$ supports $\Sigma Y_j(w)$ in x;
(c) $p = \Sigma p_i^L$.

A state vector (x, p_i, p) is called a *public equilibrium* for the economy E_2^* if (see fig. 5.2.1):
(a) the choice set $M(x)$ supports $^*C_i(x)$ in p_i, for each i;
(b) the hyperplane $H(x)$ supports $\mathring{\cap}\, ^*Y_j(w)$ in p;
(c) $p = \Sigma \lambda_i p_i$.

Before analyzing the properties of a public equilibrium, the relation between the two equilibrium concepts is established by the following property:

Property 5.2.1.

Let the economy E_2 have a valuation representation E_2^*, and let the recession cones of the social production set $Y := \Sigma Y_j(w)$ and of the social consumption set $X := \cap X_i$ be such that $0^+ Y \cap 0^+ X = \{0\}$ and $\text{Int } Y \cap X \neq \phi$. Then:
1. A Lindahl equilibrium is equivalent to a public equilibrium.
2. An allocation x is Pareto optimal in E_2 if and only if there exist an income distribution $\{\tilde{\lambda}_i\}$ such that
   (a) the choice set $M(x)$ supports $^*C_i(x)$ in p_i, for each $i \in H$;
   (b) the hyperplane $H(x)$ supports Y^* in p;
   (c) $p = \Sigma \tilde{\lambda}_i p_i$.
3. The equilibrium prices are related by $p_i^L = \lambda_i p_i$.

Proof: The economies E_2 or E_2^* meet the conditions of properties 5.1.1 and 5.1.2 and are thus representations of each other. Property 9.4.3.2 can be applied, from which it follows that (in the case of a public equilibrium): $H(x)$ supportingly separates $C^*(x) := \Sigma \lambda_i C_i^*(x)$ and $Y^* := \mathring{\cap}\, ^*Y_j(w)$ in p, if and only if:
$H(p)$ supportingly separates $C(x) = \mathring{\cap} \lambda_i^{-1} C_i(x)$ and $Y = \Sigma Y_j(w)$ in x.

The first condition gives a public equilibrium. The second condition can be rewritten as:
(a) $M(p_i)$ supports $C_i(x)$ in x, for each i;

(b) $H(p)$ supports Y in x;
(c) $p = \Sigma\lambda_i p_i$.
Substitution of $p_i = p_i^L/\lambda_i$ in the conditions (a) and (c) gives conditions of a Lindahl equilibrium, because $M(p_i^L/\lambda_i) = M(p_i^L;\lambda_i)$.

The property under 2, finally, follows from the application of the duality operation, implying that there exists a distribution $\{\tilde{\lambda}_i\}$ such that $M(p_i)$ supports $C_i(x)$ in x, for each i, $H(p)$ supports Y in x, and $p = \Sigma\tilde{\lambda}_i p_i$. See also properties 5.1.3 and 5.4.2. □

Fig. 5.2.1. A public equilibrium, with $\lambda_1 = \lambda_2 = 0.5$

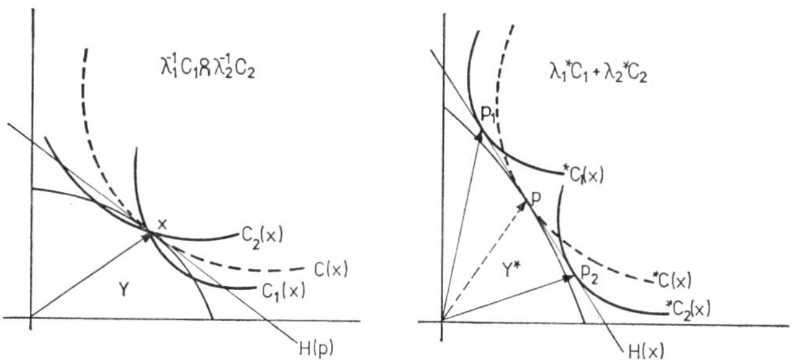

The behavior of economic agents is supposed to be such that each consumer chooses the best element in the choice set $M(x) := \{p_i \in P_i | p_i x \leqq 1\}$ according to the preference relation $\lesssim_i^*$ on P_i. If the preference relation $\lesssim_i^*$ on P_i is related with public goods, it will be called a *priority relation* on the set of public goods prices, instead of a costing relation which is defined on the set of private goods prices. For example, let $\bar{x} := \{1$ public transport; 1 public road; 1 the police service$\}$, each unit being defined in terms of quantities for the consumers (compare also table 1.2.1). Let the production prices, i.e. the marginal expansion, be given by $\bar{q} = \{0.5;\ 0.3;\ 0.2\}$.

If the weighted valuations of consumers are equal to

$$\bar{p} = \sum \lambda_i \bar{p}_i = \{0.2;\ 0.5;\ 0.3\},$$

then $\bar{p} - \bar{q} = \{-0.3;\ +0.2;\ +0.1\}$,
implying that contraction of the public transport sector and expansion of the private transport sector should be pursued in order to obtain an equilibrium. Probably, this starts an iterative process (see chapter 6). If, however, $\bar{p} = \{0.7;\ 0.1;\ 0.2\}$ because fuel has been rationed or private driving has become too expensive or dangerous, then the reverse trend

must be pursued. Evidently, if $\bar{p} = \bar{q}$ then $\bar{x}$ is to be considered an equilibrium state of public goods.

The rationality of consumers in the case of public goods can be compared with the rationality of voters. It is, of course, possible to over-emphasis certain public goods and to refuse to reveal individual valuations. But over-reporting some prices implies under-reporting other prices. Since consumers do not know a priori how the other agents will vote, their best strategy is true revelation of their preferences.

The following definitions give some precision to these ideas.

The price multifunction $p_i : X_i \to P_i$ was defined in section 4.1 by

$$p_i(x) := \{p_i \in P_i | p_i \text{ is a maximal element of } M(x) \text{ for } \lesssim_i^*\}.$$

The *social benefit multifunction* $p : X \to P$, where $X := \cap X_i$ and $P := \Sigma \lambda_i P_i$, is defined by:

$$p(x) := \sum \lambda_i p_i(x).$$

The *social cost multifunction* $q : Y \to Y^*$, where $Y := \Sigma Y_j(w)$ and $Y^* := \mathring{\cap}^* Y_j(w)$, is defined by

$$q(x) := \{\bar{q} \in Y^* | \bar{q}x = \max px, \quad \text{for} \quad p \in Y^*\}.$$

The *excess benefit multifunction* $r : X \cap Y \to P - Y^*$ is defined by:

$$r(x) := p(x) - q(x).$$

Therefore, the allocation x in E_2 is a public equilibrium if and only if $r(x) = \{0\}$.

Next, let the economy E_2 be specified by the following assumptions:

Assumptions 5.2.

On consumers (see also assumptions 4.1):

1. The consumption sets $X_i = \text{Dom}\,(C_i)$ are convex;
2. $\lesssim_i$ is a complete pre-order relation on X_i;
3. The upper preference sets $C_i(x)$ are convex;
4. The indifference sets $C_i(x) \cap C_i^{-1}(x)$ are thin;
5. All choice sets in X_i and P_i have maximal elements;
6. The consumption sets' closure equals the non-negative orthant, i.e. $\text{Cl}\, X_i = R^n_+$, and $\lesssim_i$ is monotone, i.e. $x \leqq y \Rightarrow x \lesssim_i y$.

On producers (see also section 3.2):

7. The production multifunctions $Y_j : R^n_+ \to R^n$ are point-convex, point-closed, upper-bounded, and monotone increasing multifunctions, such that $Y_j(0) = R^n_-$ and $0 \in \text{Int } Y(z)$, for $z \geq 0$.

On behavior (see also section 3.7):

8. All agents choose maximal elements in their choice sets.

The following property can be deduced from properties 5.1.1–5.1.3 and section 9.2:

Property 5.2.2.

Let the economy E_2 meet the assumptions under 5.2. Then:

1. E_2^* is a valuation representation of E_2;
2. the price sets P_i are completely preordered by the convex priority relation $\precsim_i^*$; $\precsim_i^*$ is monotone and $\text{Cl } P_i = R^{n*}_+$; the indifference sets $D_i(p) \cap D^{-1}(p)$ are thin;
3. the production cost multifunctions $^*Y_j$ are point-convex, point-starred, point-compact and non-negative, if the input is non-zero.

Relative to the price multifunctions, the following can be said:

Property 5.2.3.

Let the economy E_2 meet the assumptions under 5.2. Then:

1. the price multifunctions $p_i(x)$ and the social benefit multifunction $p(x)$ are non-empty, graph-closed and point-convex; the value $p(x) \cdot x = 1$;
2. the social cost multifunction $q(y)$ is non-empty, point-convex, point-compact and u.h.c. on X; the value $q(y) \cdot y \leqq 1$.
3. the excess benefit multifunction $r(x)$ is non-empty, point-convex and graph-closed; the value $r(x) \cdot x \geqq 0$.

Proof

1. The alleged properties of the individual price multifunctions are derived from property 4.1.4. According to properties 8.2.3 and 8.3.3, they are invariant under the operations of scalar multiplication and addition. Since the weighted sum is taken, the value remains equal to 1.
2. Since $\bar{q} \in Y^*$ is a continuous real valued function (in Y) and Y^* is a constant (i.e. continuous) and compact multifunction, the multifunction $q(x)$ is point-compact and u.h.c. according to the maximum theorem (property 8.2.4). Since Y^* is compact and convex, $q(x)$ is

non-empty and convex. Since $x \in Y$ implies that $qx \leqq 1$, for all $q \in Y^*$, it follows that $q(x) \cdot x \leqq 1$.

3. From properties 8.2.3 and 8.3.3. □

The value $r(x) \cdot x = 0$ if and only if x is an efficient production, i.e. $x \in \text{Bnd } Y$. Therefore, if $r(x) \cdot x > 0$, then $x \in \text{Int } Y$ and x is neither efficient nor Pareto optimal (see fig. 5.2.2). An efficient production does not, of course, imply a public equilibrium (see fig. 5.2.3). The excess benefit vector $r(x)$ indicates the direction in which the commodity bundle should change in order to minimize the value of $r(x) \cdot x$ and to come closer to a public equilibrium.

Fig. 5.2.2. An inefficient bundle x

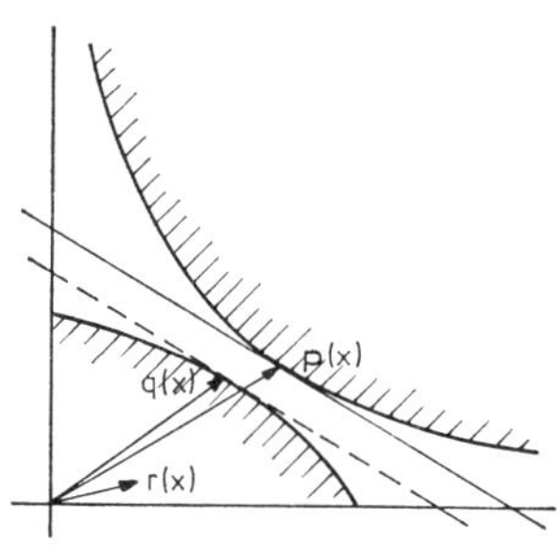

Fig. 5.2.3. An efficient disequilibrium x

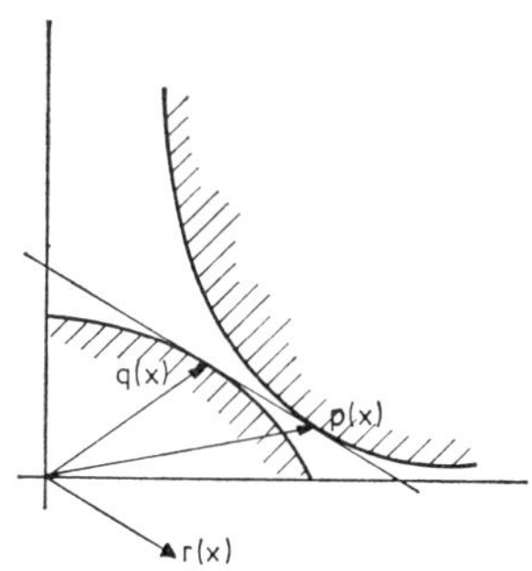

The economy E_2, its definitions and its assumptions are consistent, as is shown by the following property:

Property 5.2.4.

If the economy E_2 satisfies the assumptions under 5.2, then a public equilibrium exists.

Proof

The proof is based on the Kakutani fixed point theorem (see property 10.2.1). In order to apply this theorem, sets must be constructed which meet the conditions of the theorem. This is done as follows: Choose $\bar{z}_i \in \text{Int } \lambda_i Y \cap X_i$. This bundle $\bar{z}_i$ exists as $\lambda_i > 0$, $0 \in \text{Int } Y$ and $\text{Cl } X_\iota = R^n_+$. Then for all $z \in C_i^{-1}(\bar{z}_i)$, $p_i(z) \notin \lambda_i^{-1} Y^*$, or $\lambda_i p_i(z) \notin Y^*$, implying that $p(z) = \Sigma \lambda_i p_i(z) \notin Y^*$.

The set $\bar{X}_i := C_i(\bar{z}_i) \cap Y$ is compact, convex and non-empty (see fig. 5.2.4), and contains all equilibrium allocations.

Fig. 5.2.4.

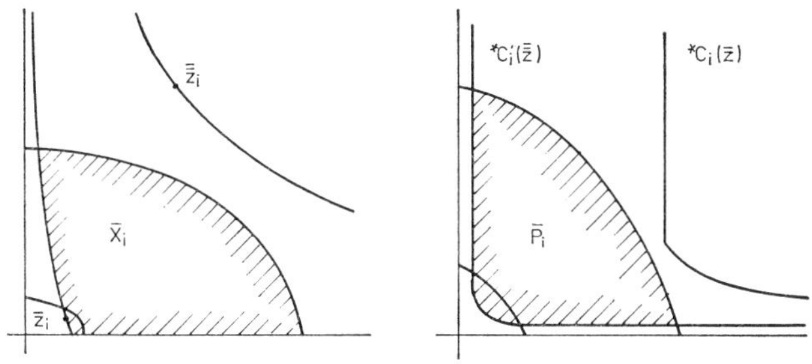

Next, choose $\bar{\bar{z}}_i \in X_i$ such that $C_i(\bar{\bar{z}}_i) \cap Y = \phi$. Thus a feasible allocation must belong to $C_i^{-1}(\bar{\bar{z}}_i)$. Since $z \in \operatorname{Int} C_i(\bar{\bar{z}}_i)$ implies that $p_i(z) \notin {}^*C_i(\bar{\bar{z}}_i)$, it follows that z is feasible only if $p_i(z) \in \operatorname{Int} {}^*C_i(\bar{\bar{z}}_i)$.

Since $p_i(z) \notin \lambda_i^{-1} Y^*$ implies that $p(z) = \Sigma \lambda_i p_i(z) \notin Y^*$ and therefore that z is not an equilibrium allocation, it follows that z is optimal only if $p_i(z) \in \lambda_i^{-1} Y^*$. The set $\bar{P}_i := {}^*C_i(\bar{\bar{z}}_i) \cap \lambda^{-1} Y^*$ is compact and convex, as ${}^*C_i(\bar{\bar{z}}_i)$ and $\lambda^{-1} Y^*$ are closed and convex, and λ^{-1} and Y^* are bounded (see property 5.2.2); the set $\bar{P}_i$ is non-empty, as $Y \cap C_i(\bar{\bar{z}}) = \phi$ implies that $\operatorname{Int} Y^* \cap {}^*C_i(z) \neq \phi$ (see property 9.4.4) and contains all equilibrium prices.

The restrictions $\bar{P}_i : X_i \to \bar{P}_i$ of the price multifunctions $p_i : X_i \to P_i$ are non-empty, point-closed, point-convex and u.h.c. (see properties 5.2.3.1 and 8.2.2.2), and so are (see property 5.2.3):
$\bar{p} : \bar{X} \to \bar{P}$, where $\bar{X} = \cap \bar{X}_i$ and $\bar{P} = \Sigma \lambda_i \bar{P}_i$;
$\bar{q} : \bar{X} \to Y^*$, and
$\bar{r} : \bar{X} \to (\bar{P} - Y^*)$.
Define the multifunction $\bar{z} : (\bar{P} - Y^*) \to \bar{X}$ by:

$$\bar{z}(r) := \{x \in \bar{X} \mid rx = \min ry, \text{ for } y \in \bar{X}\}.$$

Then $\bar{z}$ has the same properties as $\bar{q}$, as the same reasoning can be applied (see 5.2.3.2).

Finally, define the multifunction w from $(\bar{P} - Y^*) \times \bar{X}$ into itself, by $w(r, x) := \bar{r}(x) \times \bar{z}(r)$. The set $(\bar{P} - Y^*) \times \bar{X}$ is non-empty, compact and convex because both $(\bar{P} - Y^*)$ and $\bar{X}$ are also non-empty, compact and convex (see properties 8.2.3 and 8.3.3). The multifunction w is non-empty, point-closed, point-convex and u.h.c. because $\bar{r}$ and $\bar{z}$ are also. Therefore all conditions of Kakutani's fixed point theorem (property 10.2.1) are met, and w has a fixed point $(\tilde{r}, \tilde{x}) \in w(\tilde{r}, \tilde{x}) = \bar{r}(\tilde{x}) \times \bar{z}(\tilde{r})$.

Since $\tilde{x} \in \bar{z}(\tilde{r})$ implies $\tilde{r}x \geqq \tilde{r}\tilde{x}$ for all $x \in \bar{X}$ and $\tilde{r} \in \bar{r}(\tilde{x})$ implies $\tilde{r}\tilde{x} \geqq 0$ (by property 5.2.3.3), then $\tilde{r}x \geqq 0$, for all $x \in \bar{X}$, or $\tilde{r} \in \bar{X}_+^0 \supseteq R_+^{n*}$. If $\tilde{r} \in \bar{r}(\tilde{x})$ such that $\tilde{r} \ngeqq 0$, then $\tilde{x} \in \text{Bnd}\, C_i(\bar{z}_i) \cap \bar{X}$, which cannot be an equilibrium by construction. If $\tilde{r} \in \bar{r}(\tilde{x})$ such that $\tilde{r} > 0$, then $q(\tilde{x}) \cdot \tilde{x} < 1$ implying that $\tilde{x} \in \text{Int}\; Y$ and is thus inefficient and not an equilibrium. It follows that $\tilde{r} \in \bar{r}(\tilde{x})$ such that $\tilde{r} = 0$. Therefore, there exists an $\tilde{x} \in \bar{X}$ such that for some $\tilde{p}_i \in \bar{p}_i(\tilde{x})$ and $\tilde{q} \in \bar{q}(\tilde{x}) : \Sigma \lambda_i \tilde{p}_i = \tilde{q}$.

Finally, it will be shown that the state vector $(\tilde{x}, \tilde{p}_i, \tilde{q})$ is a public equilibrium for E_2. This is true if $\bar{p}_i(\tilde{x}_i) \subseteq p_i(\tilde{x}_i)$ and $\bar{q}(\tilde{x}) \subseteq q(\tilde{x})$. Since $\bar{p}_i(\tilde{x}) = M(\tilde{x}) \cap C_i^*(\tilde{x}) \cap \lambda^{-1} Y^*$, for $\tilde{x} \in \bar{X}_i \subseteq X_i$ and $p_i(\tilde{x}) = M(\tilde{x}) \cap C_i^*(\tilde{x})$, the first inclusion follows. Therefore, the choice set $M(\tilde{x})$ supports each $C_i^*(\tilde{x})$ in p_i. Since $\bar{q}(\tilde{x}) = M(\tilde{x}) \cap Y^*$, for $\tilde{x} \in \bar{X}_i \subseteq Y$ and $q(\tilde{x}) = M(\tilde{x}) \cap Y^*$, the second inclusion follows. Thus the hyperplane $H(\tilde{x})$ supports Y^* in $\tilde{q}$. Since $\tilde{q} = \Sigma \lambda_i \tilde{p}_i$, the conditions for a public equilibrium are satisfied. □

Finally, it should be noticed that the first proof of the existence of a Lindahl equilibrium for an economy with only public goods has been given by Milleron (1969). He applied a theorem of Debreu (1959) to an economy constructed in the dual space, under standard assumptions except for a priori given production prices.

Independently, Ruys (1970) has used the same idea to prove the existence of a public equilibrium for a more general economy. In his 1971 and 1972 papers, a simpler proof was substituted for the application of Debreu's theorem, using the properties of polarity operations.

5.3. EQUILIBRIUM IN AN ECONOMY WITH PRIVATE AND PUBLIC GOODS

The Lindahl equilibrium was defined in the previous section for the economy E_2 with public goods only. Although it is formally correct to do so, this certainly does not accord with the idea that a Lindahl equilibrium is an extension of the concept of a competitive equilibrium in an economy with private goods. In this section, an economy will be defined with both private and public goods, and an appropriate definition of a Lindahl equilibrium will be given.

The operations used in obtaining the social production set and the social upper-preference set in the equilibria defined thus far, are dependent on the characteristics of the commodities. If the inputs are private goods, then addition is the appropriate operation for aggregation, and convex intersection is suitable for upper-preference sets whenever public goods are considered. This is true if the economic goods are represented in the impersonal commodity space, which is allowed according to property 4.2.2.

In an economy with both private and public goods, two operations are necessary for aggregation. This creates awkward mathematical problems which should be avoided, if possible. The way out of these problems is to define the economy in the personalized commodity space, rather than in the impersonal commodity space (see section 4.2). Two approaches are possible, dual to each other.

Firstly, Foley (1970) applied the device of personalizing the public goods subspace to establish the existence of a Lindahl equilibrium. This enabled him to formally consider all commodities as private goods and apply a well known theorem of Debreu (1962). The other approach is to personalize the subspace of private goods and to formally treat all commodities as public goods, for which all consumers have a price. This idea was originally expressed by Arrow (1970) and has been used by Bergstrom (1971) to generalize the concept of a Lindahl equilibrium.

Both approaches will be shown here, and the last approach will be used to prove the existence of a Lindahl equilibrium for an economy meeting the assumptions under 5.2 (see property 5.3.3).

Let E_3 be an economy with m private goods and n public goods. There are h consumers each having a characteristic consumption structure and income fraction, and f producers each having a production multifunction defined on the subspace of public goods, such that the bundle of initial public goods, $w_N \in R^n$, determines the set of production possibilities. The bundle of initial private goods, $w_M \in R^m$, is given separately:

$$E_3 := \{H, \{X_i, \lesssim_i, \lambda_i\}; F, \{Y_j\}; (w_M, w_N)\}.$$

The allocation $\{(x_{Mi}, x_N), y_j\}$ and $\{(p_M, p_{Ni}^L), q\}$ of quantities and prices in E_3 and its polar economy E_3^* is said to be a *Lindahl equilibrium* for E_3 if:

(a) the choice sets $M[(p_M, p_{Ni}^L); \lambda_i]$ support $C_i(x_{Mi}, x_N)$ in (x_{Mi}, x_N), for each i;

(b) the hyperplane $H(q)$ supports $Y := \Sigma Y_j$ in y;

(c) $y = (\Sigma x_{Mi}, x_N)$ and
$q = (p_M, \Sigma p_{Ni}^L)$.

For every consumer i, therefore, (x_{Mi}, x_N) is a maximal element for $\lesssim_i$ of his choice set M_i, and producers are assumed to maximize their profits, i.e. to work efficiently.

Although public goods in the economy are supplied to both consumers and producers, only consumers are considered in determining the benefit price of public goods. When producers need a public good in order to produce a bundle of consumption goods more efficiently, they will decide to buy or make such a public good. Since this good will also be available for consumers, the 'free supply' will lower the benefit price for them if the commodity so supplied was at all desired. This approach of neglecting public goods necessary for production expresses the attempt of consumers to decide about the final allocation, while it does not preclude such extra public goods-inputs if efficiency criteria allow for them.

A different approach is followed by Fabre-Sender (1969) and Milleron (1972), who explicitly introduce these inputs – and their prices – in their definition of a Lindahl equilibrium.

It should also be noted that public goods needed by producers for an efficient production which have a negative benefit for consumers (such as pollution) can be considered as inputs in the production process. The opposite of these inputs (e.g. clean air) is the public good produced by choosing another – more expensive – production process and consumed by consumers, just as the opposite of labor-time is equal to the consumption good 'leisure-time'.

Further, some institutional specification about the income distribution $\{\lambda_i\}$ in E_3 should be made. If one wants to base the behavior of producers on 'individual rationality' (see section 1.4), and introduce a system of private ownership in E_3, then the resources and property shares of enterprises should be divided over consumers. Each income λ_i is then equal to the value of resources and shares possessed by consumer i, and depends on the prices to be determined. This dependence is not essential for the analysis.

The character of public goods implies that the upper-preference sets of consumers cannot simply be added, as is the case when all commodities are private goods. However, by personalizing the commodity space, it is possible to define an 'extended economy' in which the operation of addition can be applied to all relevant sets, even if the commodities are public goods. By means of this device, a public equilibrium can be formally represented by a competitive equilibrium in the extended economy.

This is shown for the economy E_2 for public goods only, defined in the previous section.

Consider the following economy:
$E_2^N := \{H, (X_i^N, \precsim_i^N, \lambda_i); Y^N\}$, where
$X_i^N := \{(0, \ldots, x_i, \ldots, 0) \in R^{hn} \mid x_i \in X_i\}$ and
$Y^N := \{(x_1, \ldots, x_i, \ldots, x_h) \in R^{hn} \mid \forall i \in H : x_i \leqq x \text{ and } x \in \Sigma Y_j(w)\}$.
The economy E_2^N is said to be a *public goods extension of* E_2. From the definition of a public good given in section 4.2, it follows that in an economy with public goods only, defined in a personalized commodity space, the individual consumption sets have at most the origin in common. This is also true for the sets X_i^N in E_2^N. The relation between $\precsim_i^N$ in X_i^N and $\precsim_i^{N*}$ in P_i^N is given in section 4.2 and in fig. 4.2.3.

The following property of a public equilibrium in E_2 can be established:

Property 5.3.1.

Let $(x, \{p_i\}, p)$ be a state in E_2 and $(\{x_i^N\}, x^N, p^N)$ be a state in E_2^N such that:

$$\text{Proj}_i\, x_i^N = \text{Proj}_i\, x^N = x \text{ and } \text{Proj}_i\, p^N = \lambda_i p_i, \text{ for all } i \in H.$$

Then $(x, \{p_i\}, p)$ is a public equilibrium in E_2 if, and only if, $(\{x_i^N\}, x^N, p^N)$ is formally a competitive equilibrium in E_2^N.

Fig. 5.3.1. A competitive equilibrium for public goods in E_2^N.

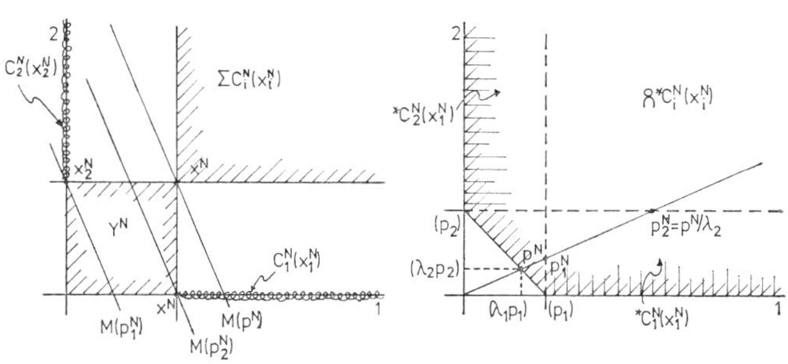

Proof
If $(x, \{p_i\}, p)$ is a public equilibrium in E_2, then:
(a) $M(p_i)$ supports $C_i(x)$ in x;
(b) $H(p)$ supports $Y := \Sigma Y_j(w)$ in x;
(c) $p = \Sigma \lambda_i p_i$.

Choose x^N, p_i^N and p^N in E_2^N, resp. in E_2^{N*}, such that the conditions above are met. Then:
(c′) $x^N = \Sigma x_i^N \in Y^N$.
Further, since for all $y^N \in Y^N$ it is true that $p^N y^N = (\lambda_1 p_1, \ldots, \lambda_h p_h) \cdot (y_1, \ldots, y_h) = \Sigma \lambda_i p_i y_i = py \leqq 1$, and this equality is also true for x^N, it follows that (b′) $H(p_N)$ supports Y_N in x_N.

Finally, since for all $y_i^N \in C_i^N(x^N)$ it is true that $p^N y_i^N = (\lambda_1 p_1, \ldots, \lambda_h p_h) \cdot (0, \ldots, y_i, \ldots, 0) = \lambda_i p_i y_i \geqq \lambda_i$, given (a) above, and this equality is true for x^N, it follows that
(a′) $M(p^N; \lambda_i)$ supports $C_i^N(x^N)$ in x^N, for all i.
The conditions (a′), (b′) and (c′) imply that $(\{x_i^N\}, x^N, p^N)$ is a competitive equilibrium.

The converse statement follows immediately from the reduction of E_2^N into E_2 obtained by projecting the sets in R^{hn} into R^n. □

The 'competition' in E_2^N has no economic significance because exchange of the personalized commodities x_i^N will (and can) never take place. Only its formal correspondence with the definition of a competitive equilibrium is relevant.

Next, the device of personalizing private goods will be shown to permit use of the operation of convex intersection of all upper-preference sets, which operation has been used for a public goods-only economy (see section 5.2). Thus, a competitive equilibrium can be formally represented by a 'public equilibrium', as will be shown for the economy E_1 with private goods only.

Consider the following economy:
$E_1^M := \{H, (X_i^M, \precsim_i^M, \lambda_i); Y^M\}$, where
$X_i^M := X_1 \times \ldots \times X_i \times \ldots \times X_h$, on which set the preference relation $\precsim_i^M$ is defined by:
$(x_i^1, \ldots, x_i^1, \ldots, x_h^1) \precsim_i^M (x_1^2, \ldots, x_i^2, \ldots, x_h^2)$ if and only if $x_i^1 \precsim_i x_i^2$, for $\precsim_i$ on X_i;
$Y^M := \{(x_1, \ldots, x_h) \in R^{hm} \mid \Sigma x_i = x \text{ and } x \in Y\}$.

The economy E_1^M is said to be a *private goods extension of* E_1. Compare also section 4.2 and fig. 4.2.1 on private goods. The following property of a competitive equilibrium can be established.

Property 5.3.2.

Let $(\{x_i\}, x, p)$ be a state in E_1 and $(x_M, \{p_i^M\}, p^M)$ be a state in E_1^M such that:

$$\text{Proj}_i\, x^M = x_i, \ \text{Proj}_i\, p_i^M = p_i \text{ and } \text{Proj}_i\, p^M = p = \lambda_i p_i,$$

for each $i \in H$. Then $(\{x_i\}, x, p)$ is a competitive equilibrium in E_1 if, and only if, $(x^M, \{p_i^M\}, p^M)$ is formally a public equilibrium in E_1^M.

Fig. 5.3.2. A public equilibrium for private goods in E_1^M

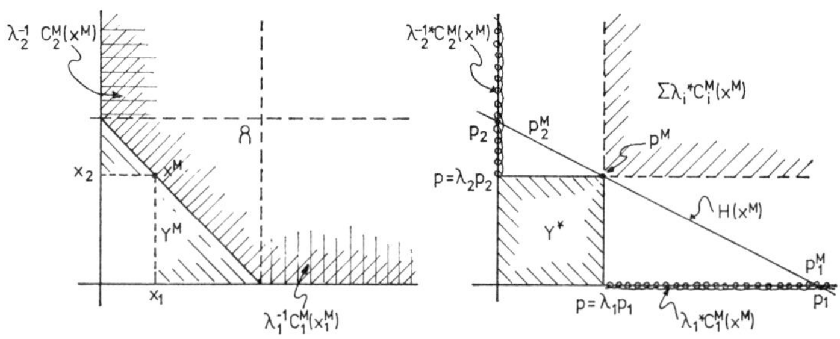

Proof

If (x_i, x, p) is a competitive equilibrium in E_1, then:
(a) $M(p; \lambda_i)$ supports $C_i(x_i)$ in x_i;
(b) $H(p)$ supports Y in x, and
(c) $x = \Sigma x_i$.
Choose x^M, $\{p_i^M\}$ and p^M such that the above conditions and those outlined in the definition of E_1^M are met. Since $p_i^M \in C_i^{M*}(x^M)$, and $p_i^M \cdot p_j^M = 0$, for $i \neq j$, it follows that
(c′) $\Sigma p_i^M = p^M$.

For every $y^M \in Y^M$ it is true that

$p^M y^M = (p, \ldots, p) \cdot (y_1, \ldots, y_h) = \Sigma p y_i = py \leqq 1$, with equality for $y^M = x^M$. Thus:
(b′) $H(p^M)$ supports Y^M in x^M.

Finally, for any given $i \in H$ it holds that $y^M \in C_i^M(x^M)$ implies

$p_i^M y^M = (0, \ldots, p_i, \ldots, 0) \cdot (y_1, \ldots, y_h) = p_i y_i = \lambda_i^{-1} p y_i \geqq 1$,
with equality for $y^M = x^M$. Therefore
(a′) $M(p_i^M)$ supports $C_i^M(x^M)$ in x^M, for each i.

Conditions (a′), (b′) and (c′) imply that $(x^M, \{p_i^M\}, p^M)$ is a public equilibrium in E_1^M. The converse statement is easily verified by projecting the personalized commodity space into R^m (see fig. 5.3.3). □

In fig. 5.3.3, $x_1 = 1/3x$ and $x_2 = 2/3x$; this corresponds with $p_1 = 3p$ and $p_2 = 3/2p$, or $p = 1/3p_1 = 2/3p_2$. The scalars, λ_i, thus represent the income distribution and determine the allocation of x over the individual consumers, just as in economy E_1^M.

Fig. 5.3.3. A competitive equilibrium in E_1

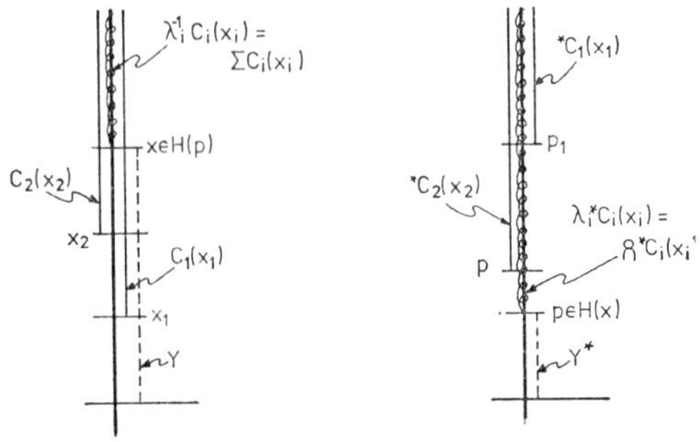

It can be repeated that the equilibrium in E_1^M is called 'public' for mathematical reasons only. The personalized commodities are not public, as they do not meet the conditions of public goods. It is now possible, however, to prove the existence of a Lindahl equilibrium for the economy E_3 with both public and private goods in a rather simple way, because the same operations of aggregation can be applied to both private public and goods.

Consider *the private goods extension* E_3^M of E_3, defined by:

$$E_3^M := \{H, (X_i^M, \precsim_i^M, \lambda_i);\ F, (Y_j^M);\ w^M \in R^{hm+n}\}, \text{ where}$$
$$X_i^M := X_{M1} \times \ldots \times X_{Mh} \times X_{Ni},$$

X_{Mi}, resp. X_{Ni} being the image of X_i under the projection $R^{m+n} \to R^m$, resp. $R^{m+n} \to R^n$. On X_i^M the preference ordering $\precsim_i^M$ is defined as follows:

$$(x_{M1}^1, \ldots, x_{Mh}^1, x_{Ni}^1) \precsim_i^M (x_{M1}^2, \ldots, x_{Mh}^2, x_{Ni}^2)$$

if and only if $(x_{Mi}^1, x_{Ni}^1) \precsim_i (x_{Mi}^2, x_{Ni}^2)$ for $\precsim_i$ on X_i, for every $i \in H$.

$$Y_j^M(w_N) := \{(x_{M1}, \ldots, x_{Mh}, x_N) \in R^{hm+n} \mid (\Sigma x_{Mi}, x_N) \in Y_j(w_N)\}.$$

Property 5.3.3.

If the economy E_3 satisfies the assumptions under 5.2, then a Lindahl equilibrium exists in E_3.

Proof

Construct the private goods extension E_3^M. This economy is formally equal to E_2. Since the assumptions 5.2 are equally valid in E_3^M, as can be easily seen, property 5.2.4 can be applied. Thus there exists a public equilibrium $(x^M, \{p_i^M\}, p^M)$ in E_3^M. Since $p_i^M \in C_i^{M*}(x^M)$ and $p^M = \Sigma\lambda_i p_i^M$, it follows that:

$$x^M = (x_{M1}, \ldots, x_{Mh}, x_N),\ p_i^M = (0, \ldots, p_{Mi}, \ldots, 0, p_{Ni}) \quad \text{and}$$
$$p^M = (p_M, \ldots, p_M, p_N) = (\lambda_1 p_{M1}, \ldots, \lambda_h p_{Mh}, \Sigma\lambda_i p_{Ni}).$$

Further, the Lindahl price for public goods is defined by $p_{Ni}^L := \lambda_i p_{Ni}$.

Projection of E_3^M into the impersonal commodity space R^{m+n} gives the following results:
$M(p_i^M)$ supporting $C_i^M(x^M)$ in x^M implies that
(a) $M(p_{Mi}, p_{Ni}) = M[(p_M, p_{Ni}^L); \lambda_i]$ supports $C_i(x_{Mi}, x_N)$ in (x_{Mi}, x_N), for each $i \in H$.

Since $H(p^M)$ supports $\Sigma Y_j^M(w_N)$ in $\Sigma y_j^M := x^M$, and $p^M x^M = p_M \Sigma x_{Mi} + p_N x_N = p_M x_M + p_N x_N$, it follows that
(b) $H(p_M, p_N)$ supports $\Sigma Y_j(w_N)$ in (x_M, x_N).

Finally, from $p^M = \Sigma\lambda_i p_i^M$ and the definition of $Y_j^M(w_N)$ it follows that
(c) $p_N = \Sigma\lambda_i p_{Ni} = \Sigma p_{Ni}^L$ and $x_M = \Sigma x_{Mi}$.

Thus the allocation $\{(x_{Mi}, x_N), y_j\}$ and $\{(p_M, p_{Ni}^L), (p_M, p_N)\}$, which exists, satisfies the conditions of a Lindahl equilibrium in E_3. □

As a corolary of this property it follows that the economy E_1 has a competitive equilibrium when the assumptions 5.2 are satisfied. This corolary can also be deduced from properties 5.3.2 and 5.2.4:

Property 5.3.4.

If the economy E_1 satisfies the assumptions under 5.2, then there exists a competitive equilibrium in E_1.

Finally, it should be mentioned that Roberts (1972) has shown the existence of a Lindahl equilibrium for an economy with a continuum of agents.

5.4. A TWO-LEVEL PRICE EQUILIBRIUM

The duality approach is not at all essential for the theory underlaying a Lindahl equilibrium, as can be deduced from the previous section. The device of extending the commodity space permits the application of tools which are constrained to the commodity space. The only advantage of the duality approach in the Lindahl theory is to be found in mathematical arguments.

It has been stressed in section 5.2 that the definition of a public equilibrium depends directly on instruments supplied by the duality theory. The agents have choice sets in the price space and are supposed to choose the maximal price in those sets. Since no relation exists with a private commodity, and the direct influence of some agents' decision on his own well-being is infinitesimal, this assumption on the behavior of agents is not too restrictive. Examples for which this approach makes sense are a (city-) council or a parliament which have to decide about the composition of a bundle of public goods, under the restriction that the amount to be spent is given and that the decision have no influence on the level of private goods and their prices; i.e., when the economy can be assumed to contain only public goods.

In this section an attempt will be made to develop an equilibrium concept in which decisions are made on two levels separately: the level of public goods and the level of private goods. An income distribution is given or determined for each level, generating individual budget sets for private goods and fixing individual weights for the price set of public goods. These distributions may be equal, but need not necessarily be. The composition of allocation between the two levels and within each level determines the production prices. These production prices can be used to determine whether the composition within each level is such that the benefit prices equal the cost prices, and between both levels such that production is feasible and efficient. Such a state will be called a two-level price equilibrium.

Thus, the consumers have two distinct choice sets: one for public goods and one for private goods. These choice sets are linked through the social production set, which can be composed of individual production sets.

Consider the economy E_3 in section 5.3 and replace the income distribution $\{\lambda_i\}$ by two distributions, $\{\lambda_{Mi}\}$ and $\{\lambda_{Ni}\}$ for private goods and public goods respectively. Each distribution consists of positive fractions adding up to 1. This economy E_4 is called a *two-level economy* and is defined by:

$$E_4 := \{H, \{X_i, \lesssim_i, \lambda_{Mi}, \lambda_{Ni}\}; F, \{Y_j\}; (w_M, w_N)\}.$$

Let $A := [\{(x_{Mi}, x_N)\}, \{y_j\}, \{(p_M, p_{Ni})\}, q]$ be a state in E_4. The *share of private goods* in this allocation is defined by $\mu := p_M \Sigma x_{Mi}$, and the value of public goods is equal to $1-\mu = \Sigma p_{Ni} x_N$. Thus $0 \leqq \mu \leqq 1$. Two upper-preference sets are defined by A; one in the private goods subspace and the other in the public goods subspace:

$$C_{Mi}(x_{Mi}, x_N) := \{x \in R^m | (x, x_N) \in C_i(x_{Mi}, x_N)\}$$
$$C_{Ni}(x_{Mi}, x_N) := \{x \in R^n | (x_{Mi}, x) \in C_i(x_{Mi}, x_N)\}.$$

The state A is said to be a *two-level price equilibrium* in E_4 if:

(a–1) the allocation at the level of private goods is a competitive equilibrium;

(a–2) the allocation at the level of public goods is a public equilibrium;

(b) the allocation of inputs in production between the two levels maximizes profits of the production sector;

(c) the prices of outputs are equal to market prices for private goods and to social benefit prices for public goods;
the social supply of commodities equals social demand.

Formally, if:

(a–1) the choice sets $M(p_M;\ \mu\lambda_{Mi})$ support $C_{Mi}(x_{Mi},\ x_N)$ in x_{Mi}, for all $i \in H$, given $\mu = 1 - p_N x_N$;

(a–2) the choice sets $M(x_N;\ 1-\mu)$ support $C^*_{Ni}(x_{Mi},\ x_N)$ in p_{Ni}, for all $i \in H$, given $\mu = \Sigma p_M x_{Mi}$;

(b) the hyperplane $H(q)$ supports $\Sigma Y_j(w_N)$ in $y := \Sigma y_j$;

(c) $q = (p_M,\ \Sigma\lambda_{Ni} p_{Ni})$ and $(\Sigma x_{Mi},\ x_N) = y + (w_M,\ w_N)$.

According to this definition, consumers choose a maximal element from their budget set of private goods, given a bundle of public goods. The budget set of the i-th consumer is determined by his weight, λ_{Mi}, in the distribution of the value of private goods, μ. Consumers also choose a maximal element from the price set of a public goods proposal, $M(x_N;\ 1-\mu)$, given their bundle of private goods.

The bundle of private goods chosen is an equilibrium choice if the given bundle of public goods has a cost price equal to the weighted average of the benefit prices, $\Sigma\lambda_{Ni} p_{Ni}$. And the chosen price of public goods is an equilibrium choice if the given bundle of private goods is a choice from the budget set of private goods.

The share of private goods in the economy, μ, and the value of public goods, $1-\mu$, is defined by these equilibrium vectors.

Just a description of an equilibrium is not enough for the description of the procedure necessary to arrive at such an sub-optimum. In a private goods economy, demand and supply are adjusted through fluctuations in the proposed market prices, and in a public goods economy, costs and benefits are adjusted through fluctuations in the public goods proposal. In the economy E_4 these mechanisms remain valid, but the share of private goods, μ, has also to be adjusted.

Since the consumption decisions are divided into two levels (note that they are not independent), and production decisions are made according to the same criterion, viz. efficiency or profit, adjustments of the share of private goods must be made through the production sector. Such a function will be defined below.

A solution similar to the two-level price equilibrium has been proposed by Drèze (1974) for an investment model under private ownership, in which uncertain events are (local) public goods for the owners of a firm.

A two-level price equilibrium is more realistic than a Lindahl equilibrium, for the following reasons:

(a) it permits separation of decisions according to the level of external effects of the commodities considered;

(b) it permits 'weight' or 'income' distributions for consumers which are different at each level;

(c) all choice sets do not depend directly on the individual chosing.

It is quite evident that in most societies economic decisions are organized in such a way that all these three points are realized.

The cost of this approach, however, is that a two-level price equilibrium need not be Pareto optimal. It belongs thus to the class of 'second best solutions'. Pareto optimality can be obtained under special conditions, as mentioned in properties 5.4.3 and 5.4.4.

An allocation of quantities is Pareto optimal if and only if all agents individually *and* aggregated are in a state of equilibrium. In that case there exists a hyperplane supportingly separating the suitably aggregated production sets and upper-preference sets.

A two-level price equilibrium A in E_4 implies a state of equilibrium for all individual agents and for the production sector, but not necessarily

Fig. 5.4.1. A two-level price equilibrium A which is not Pareto optimal

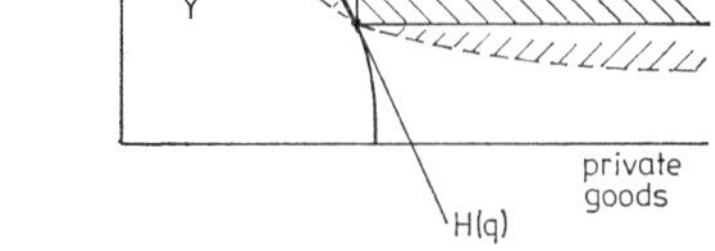

for the consumption sector, which may benefit from a shift in the shares given to the levels of private and public goods (see fig. 5.4.1).

In order to show this, the class of economies $\{E_4\}$ will be somewhat restricted to obtain an equivalent definition of a two-level price equilibrium in terms of supportingly separating hyperplanes in the quantity space.

Property 5.4.1.

Let the economy E_4 satisfy the following conditions:

1. The upper-preference sets $C_i(x)$ are aureole-reflexive for each $i \in H$ and $x \in X_i$, i.e. they are closed, convex, aureoled and do not contain 0.
2. The production sets $Y_j(w)$ are star-reflexive for each $j \in F$ and $w \in \mathrm{Dom}(Y_j)$, i.e. they are closed, convex and contain 0.
3. The social production set $Y(w)$ and the social consumption set X have an interior point in common and their recession cones have only the origin in common.

Then the state $A := [\{(x_{Mi}, x_N)\}, \{y_j\}, \{(p_M, p_{Ni})\}, q]$ is a two-level price equilibrium for E_4 if and only if the hyperplane $H(q)$ supportingly separates the sets $\{\Sigma C_{Mi}(x_{Mi}, x_N) \times \bigcap \lambda_{Ni}^{-1} C_{Ni}(x_{Mi}, x_N)\}$ and $Y := \Sigma Y_j(w)$ in $(\Sigma x_{Mi}, x_N)$.

Proof

Conditions 1. and 2. imply that the polar economy E_4^* is a valuation representation of E_4, according to properties 5.1.1 and 5.1.2.

Construct the economy $E_{M4} := \{H, \{X_{Mi}, \precsim_{Mi}, \lambda_{Mi}\}; F, \{Y_j\}; w_M\}$ from E_4 by intersecting the sets X_i, resp. Y_j in R^{m+n} with the affine subset $\{(u, v) \in R^{m+n} \mid v = x_N\}$, and by calling the images of these sets under the projection on R^m, X_{Mi} and Y_{Mj}, respectively.

Conditions (a), (b) and (c) of the definition of a two-level price equilibrium are equivalent to the following condition in E_{M4} (see section 5.1): the hyperplane $H(p_M; \mu)$ supportingly separates $\Sigma C_{Mi}(x_{Mi}, x_N)$ and $Y_M := \Sigma Y_{Mj}$ in Σx_{Mi}.

Similarly, the economy E_{N4} is constructed by fixing the private goods components x_{Mi} and y_{Mj}. Then the conditions (a), (b) and (c) in E_{N4} are equivalent to (see property 5.1.4): the hyperplane $H(x; 1-\mu)$ supportingly separating the sets $\Sigma\lambda_{Ni}{}^*C_{Ni}(x_{Mi}, x_N)$ and $Y_N^* := (\Sigma Y_{Nj})^*$ in $\Sigma\lambda_{Ni}p_{Ni}$. According to property 9.4.3.2 (which can be applied, given condition 3. above), this is equivalent to the condition that the hyperplane $H(\Sigma\lambda_{Ni}p_{Ni}; 1-\mu)$ supportingly separates the sets $\bigcap \lambda_{Ni}^{-1} C_{Ni}(x_{Mi}, x_N)$ and Y_N in x_N.

Therefore, if state A is a two-level price equilibrium in E_4, then the hyperplane $H(q)=H(p_M, \Sigma\lambda_{Ni}p_{Ni})$ supports the set $\{\Sigma C_{Mi}(x_{Mi}, x_N)\times \bigcap \lambda_{Ni}^{-1} C_{Ni}(x_{Mi}, x_N)\}$ in $(\Sigma x_{Mi}, x_N)$, from the arguments given above. Further, $H(q)$ supports $\Sigma Y_j(w)$ in the same point via condition (b) of the definition. Since $Y_M\times Y_N\subseteq Y$, the hyperplane $H(q)$ supportingly separates the sets $\{\Sigma C_{Mi}(x_{Mi}, x_N)\times \bigcap \lambda_{Ni}^{-1} C_{Ni}(x_{Mi}, x_N)\}$ and Y.

For the converse, conditions (b) and (c) of the definition follow immediately, and conditions (a) after dividing the economy into E_{M4} and E_{N4} according to the rules described above. □

Necessary and sufficient conditions for Pareto optimality of an allocation in E_4 can be formulated by means of a supportingly separating hyperplane if a correct operation of aggregation is defined. Since the upper-preference sets have to be added in one direction and convex-intersection is required for the other direction, the desired operation can be compared with the operation of partial addition of sets, defined in section 8.3.

Property 5.4.2.

The allocation $A:=[\{x_{Mi}, x_N)\}, y]$ in E_4 is Pareto optimal if, and only if, there exists an income distribution $\{\tilde{\lambda}_{Ni}\}$ for public goods and there exist prices $\tilde{q}$ such that the hyperplane $H(\tilde{q})$ supportingly separates the *social upper-preference set*

$$C(A;\tilde{\lambda}_{Ni}):=\left\{(z_M, z_N)\in X \;\middle|\; \begin{array}{l}\exists(y_{Mi}, \tilde{\lambda}_{Ni}y_N)\in C_i(x_{Mi}, x_N), \text{ such that}\\ z_M=\Sigma y_{Mi} \text{ and } z_N=\tilde{\lambda}_{Ni}y_N, \text{ for all}\\ i\in H\end{array}\right\}$$

and the social production set Y in $(\Sigma x_{Mi}, x_N)=y$.

Proof

Suppose that A is not Pareto optimal. Then there exists a Pareto superior allocation A' such that for some consumer $h\in H$ it is true that $(x'_{Mh}, x'_N)\in$ Int $C_h(x_{Mh}, x_N)$, for all $i\in H: (x'_{Mi}, x'_N)\in C_i(x_{Mi}, x_N)$, and $(\Sigma x'_{Mi}, x'_N)\in Y$. It follows that Int $C(A)\cap Y\neq\phi$, which excludes the existence of a hyperplane separating $C(A)$ and Y (see property 9.1.1).

Conversely, if $H(q)$ does not supportingly separate $C(A)$ and Y, then either $H(q)$ strongly separates $C(A)$ and Y, or $H(q)$ intersects one or both sets. In the first case, the allocation A is not feasible. In the second case, there exists a Pareto superior allocation A'. □

It follows immediately from properties 5.4.1 and 5.4.2 that a two-level price equilibrium A is not necessarily Pareto optimal. From the same

properties it can be deduced that a two-level price equilibrium is Pareto optimal if, for example, the preferences for public goods and private goods are perfectly complementary, i.e. if $C(A; \lambda_{Ni}) = \{\Sigma C_{Mi}(x_{Mi}, x_N) \times \aleph \lambda^{-1} C_{Ni}(x_{Mi}, x_N)\}$. This assumption is not so realistic in a (small) neighborhood of the equilibrium allocation.
To assess the significance of the share of private goods μ, resp. public goods $(1-\mu)$, in the budget sets used in defining a two-level price equilibrium, it is interesting to formulate conditions under which this equilibrium reduces to a Lindahl equilibrium.

Property 5.4.3.

Let the economy E_4 satisfy the three conditions given in property 5.4.1.

If the income distributions in E_4 are equal for public goods and private goods, i.e. $\lambda_{Mi} = \lambda_{Ni} = \lambda_i$, for each $i \in H$, and if the share of private goods is personalized, i.e. $\mu_i \in \{p_M x_{Ni} \mid (\lambda_i^{-1} p_M, p_{Ni}) \in C_i^*(x_{Mi}, x_N)\}$, then a two-level price allocation satisfies the conditions of a Lindahl equilibrium and is thus Pareto optimal.

Proof
Conditions (a–1) and (a–2) can be replaced by: the choice sets $M(p_M; \mu_i \lambda_i)$ supporting $C_{Mi}(x_{Mi}, x_N)$ in x_{Mi}, and the choice sets $M(x_N; 1-\mu_i)$ supporting $C_{Ni}^*(x_{Mi}, x_N)$ in p_{Ni}. Due to property 9.4.3.2, the last condition is equivalent to: the choice sets $M[\lambda_i p_{Ni}; (1-\mu_i)\lambda_i]$ supporting $C_{Ni}(x_{Mi}, x_N)$ in x_N.

Since μ_i is chosen such that $(\lambda_i^{-1} p_M, p_{Ni}) \in C_i^*(x_{Mi}, x_N)$, the hyperplane $H(\lambda_i^{-1} p_M, p_{Ni}) = H[(p_M, \lambda_i p_{Ni}); \lambda_i]$ supportingly separates $\{M(p_M; \mu_i \lambda_i) \times M[\lambda_i p_{Ni}; (1-\mu_i)\lambda_i]\}$ and $C_i(x_{Mi}, x_N)$. Therefore, the budget set $M[(p_M, \lambda_i p_{Ni}); \lambda_i] = M[(p_M, p_{Ni}^L); \lambda_i]$ corresponding to the above hyperplane supports the set $C_i(x_{Mi}, x_N)$ in (x_{Mi}, x_N), for all $i \in H$.

Conditions (b) and (c) of a Lindahl equilibrium are identical with those of a two-level price equilibrium. □
From property 5.4.3 it can be deduced that if the share of private goods μ can be chosen such that there exist personal shares μ_i satisfying the condition:

$$\left(\frac{\mu}{\mu_i}\right) \lambda_{Mi} = \left(\frac{1-\mu}{1-\mu_i}\right) \lambda_{Ni} = \lambda_i, \quad \text{for all} \quad i \in H,$$

then the two-level price equilibrium is Pareto optimal:

Property 5.4.4.

Let the economy E_4 satisfy the three conditions given in property 5.4.1.

If the social share of private goods μ can be chosen such that theie exist individual shares μ_i in the equilibrium allocation satisfying the equality:

$$\frac{\mu}{1-\mu} = \left(\frac{\mu_i}{\lambda_{Mi}}\right) \Big/ \left(\frac{1-\mu_i}{\lambda_{Ni}}\right), \quad \text{for all} \quad i \in H,$$

then the two-level price equilibrium is Pareto optimal.

One method of obtaining Pareto optimality is, therefore, to give a higher weight in the private goods income distribution (or lower weight in the public goods weight distribution) to a consumer who has a relatively high individuel share of – and preference for – private goods (or low share of public goods).

Further, for every bundle x_N of public goods in the economy E_4, an equilibrium allocation in the subspace of private goods can be determined through the market mechanism. This is expressed by the multifunction $a: Y_{N+} \to Y^h_{M+}$, where $Y_{N+} := \text{Proj}_N\, Y \cap R^n_+$, and $Y_{M+} := \text{Proj}_M\, Y \cap R^m_+$:

$$a(x_N) := \left\{ \{x_{Mi}\} \,\middle|\, \begin{array}{l} \exists \bar{p}_M\colon H(\bar{p}_M) \text{ supportingly separates } f(x_N) \text{ and} \\ \sum C_{Mi}(x_{Mi}, x_N) \text{ in } \sum x_{Mi}. \end{array} \right\}$$

Similarly, given an allocation of private goods (x_{Mi}), an equilibrium bundle of public goods in the corresponding subspace can be determined through the referendum mechanism. This is expressed by the multifunction $b: Y^h_{M+} \to Y_{N+}$, defined by:

$$b(\{x_{Mi}\}) := \left\{ x_N \,\middle|\, \begin{array}{l} H(x_N) \text{ supportingly separates } {}^*f^{-1}(x_M) \\ \text{and } \sum \lambda_{Ni}\, {}^*C_{Ni}(x_{Mi}, x_N) \text{ in } \bar{p}_N. \end{array} \right\}$$

The existence of a two level equilibrium in E_4 can now be shown under rather restrictive conditions.

Property 5.4.5.

Let the economy E_4 satisfy the assumptions 5.2. Assume also that the multifunctions $a(x_N)$ and $b(\{x_{Mi}\})$ are graph-closed and point-convex. Then there exists a two level price equilibrium.

Proof
Consider the multifunction c from $Y^h_{M+} \times Y_{N+}$ into itself, defined by $c(\{x_{Mi}\}, x_N) := a(x_N) \times b(\{x_{Mi}\})$. The assumptions 5.2 imply that both multifunctions are non-empty on their respective domains and that these domains are compact, convex and non-empty. Since a and b are graph-closed, they are point-closed and u.h.c. (see property 8.2.2). The condi-

tions for the existence of a fixed point are thus fulfilled (see property 10.2.1), and there exist an allocation $x := (\{x_{Mi}\}, x_N)$ such that $\{x_{Mi}\} \in a(x_N)$ and $x_N \in b(\{x_{Mi}\})$. The prices corresponding to this allocation are $\{\bar{p}_{Ni}\}$ and $\bar{p}_M$. Choose $\mu \in (0, 1)$ and call $p_M := \mu \bar{p}_M$, and $p_{Ni} := (1-\mu)\bar{p}_{Ni}$. The state $A := [\{(x_{Mi}, x_N)\}, y, \{(p_M, p_{Ni})\}, q]$ is a two level equilibrium if $H(q)$ supports Y.

Suppose that $H(q) \cap \text{Int } Y \neq \phi$. Then $qy > 1$ and $q \notin Y^*$ (see fig. 5.4.2). Since $0 \in \text{Int } Y$, the set Y^* is compact. Further, since $q_M = p_M$ supports $Y_M = f(x_N)$ and $q_N = p_N$ support $Y_N = f^{-1}(x_M)$, and there exist a positive vector in Y^* there also exist non-negative scalars α and β such that $(\alpha q_M, \beta q_N) := \bar{q}$ maximizes $\bar{q}x$ on Y^*.

Fig. 5.4.2. Inflating public goods prices and deflating private goods prices from p to $\bar{p}$

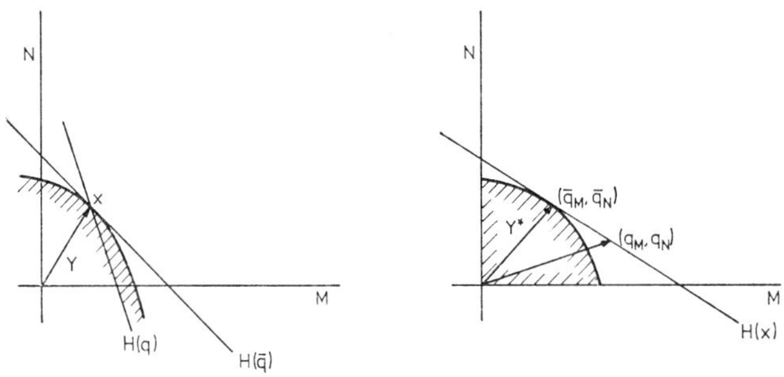

Call $\bar{\mu} := \alpha p_M x_M = \alpha\mu$ and $\bar{p}_{Ni} := \beta p_{Ni}$, for each $i \in H$. It is evident that $\bar{q}x = \alpha q_M x_M + \beta q_N x_N = \alpha\mu + \beta(1-\mu) = \bar{\mu} + (1-\bar{\mu}) = 1$. Now:
(a–1) $M(\bar{p}_M; \bar{\mu}\lambda_{Mi})$ supports $C_{Mi}(x_{Mi}, x_N)$ in x_{Mi};
(a–2) $M(x_N; 1-\bar{\mu})$ supports $C^*_{Ni}(x_{Mi}, x_N)$ in $\bar{p}_{Ni}$;
(b) $H(\bar{q})$ supports Y in $(\Sigma x_{Mi}, x_N)$, and
(c) $\bar{q} = (\bar{p}_M, \Sigma\lambda_{Ni}\bar{p}_{Ni})$.
Therefore, the allocation $[\{(x_{Mi}, x_N)\}, y, \{(\bar{p}_M, \bar{p}_{Ni})\}, \bar{q}]$, which exists, is a two-level price equilibrium. □

The proof indicates simultaneously the weakness and the strength of the concept, since equilibrium can be obtained by proportionally inflating prices at one level and proportionally deflating prices at the other level, once the equilibria on each level have been determined. Whether this equilibrium is also Pareto optimal depends on the social upper-preference set.

A two-level price equilibrium belongs to the class of second best solutions. In a problem of second best, allocations are considered optimal which are not necessarily Pareto optimal. A rather confusing debate about the theory of second best was begun by Lipsey and Lancaster (1956) on one hand, and Davis and Whinston (1965) on the other. It has been observed by Guesnerie (1973) that the relevance of second best solutions for welfare economics is increasing, because in those cases not only the physical or technological constraints are taken into account, but also non-physical constraints such as the conditions imposed by allocation mechanisms or procedures. However, a general formulation of the second best problem does not seem feasible, according to Bohm (1967, p. 314), to the extent that one wishes to obtain general results, in particular, general guidelines for optimum economic policy. Each model should therefore be considered with this in mind.

An allocation mechanism for arriving at a two-level price equilibrium is given by the referendum mechanism, defined in section 6.5.

5.5. FINANCING THE PUBLIC SECTOR

Public goods in an economy are supplied through a collective decision, but the cost of these public goods must also be shared by the agents in an economy. What is meant by a 'collective decision' in a decentralized economy has been discussed in the preceding sections. In this section a description will be given of four models each of which solves the financing problem in a characteristic way. The first solution is complete correspondence between individual benefit and individual cost, given the wealth distribution, i.e. the Lindahl approach. The second, a model of Foley's (1967) is one in which taxes are individually determined, but no use is made of the individual prices of public goods. In the other models, the direct link between individual benefits and costs at the public level is completely cut: one by Fourgeaud (1969), based on a system of national accounts of private goods with production and redistribution operations, and the model given in section 5.4 in which two kinds of prices are used so that the individual price of public goods only indicates the individual's benefit.

As long as only private goods exist in the economy, it is easy to finance the allocation in the economy. The market mechanism guarantees that somebody deciding to supply himself with a specific quantity, must at the same time pay his share in the cost of production, which share depends only on the quantity demanded. The concept of a competitive equilibrium demonstrates that social production costs exactly equal the sum of shares paid by the consumers.

The Lindahl solution is formally an extension of a competitive equilibrium. However, the share paid by a consumer also depends on the price at which the consumer valuates the public good supplied, which valuation is determined by the benefit the consumer receives from the public good. If a consumer is behaving rationally, in the sense of minimizing his cost, it is not at all sure that he will reveal the true price. Since consumers tax themselves in revealing the individual prices of public goods, it is quite probable that the prices revealed will be too low and that the public goods will not be adequately financed (if produced at all).

Foley (1967) has suggested that the government in an economy makes a public goods proposal via a system of individual taxes $\{\tau_i\}$. For an equilibrium state he requires that the public sector proposal cannot be turned down by any other through the political mechanism. Using the unanimity rule for the political mechanism, he defines a *public competitive equilibrium* as consisting of a feasible allocation $(\{x_{Mi}\}, x_N)$, a price vector p and a tax system $\{\tau_i\}$ with $p_N x_N = \Sigma\tau_i$, such that:

(a) $M(p_M; \lambda_i - \tau_i)$ supports $C_{Mi}(x_{Mi}, x_N)$ in x_{Mi}.

(b) $H(p)$ supports Y in $(\Sigma x_{Mi}, x_N)$.

(c) There is no public sector proposal $(\bar{x}_N, \{\bar{\tau}_i\})$ at prices p, so that for every i there exists $\bar{x}_{Mi}$ with $(\bar{x}_{Mi}, \bar{x}_N) \succ_i (x_{Mi}, x_N)$ and $p_M \bar{x}_{Mi} \leqq p_M x_{Mi} + \tau_i - \bar{\tau}_i$.

However, since $\{\tau_i\}$ may represent any tax system, it is formally equal to the distribution $\{\tilde{\lambda}_{Ni}\}$ mentioned in property 5.4.2. Therefore, the set of public competitive equilibria is equal to the set of Pareto optima when only the public level is considered. This has been noticed already by Milleron (1972, section II.4), who concludes that it is an interesting concept for characterizing Pareto optimal solutions without introducing personalized prices. It is left to political mechanisms, however, to determine a proposal for public goods and individual taxes that will be sustained by the collective rationality.

Under assumptions which are slightly stronger than the assumptions under 5.1, Foley has shown that there exists a public competitive equilibrium if taxes are proportional to individual wealth and if the taxation rate is uniform, i.e. if $\tau_i = (1-\mu)\lambda_i$ with $(1-\mu) = p_N x_N$. (The same symbols are used as in section 5.4, which means that the value of the bundle of resources possessed by the i-th consumer, $p_M x_{Mi}$, has been substituted for λ_i.)

With each public sector proposal $(x_N, \{\tau_i\})$ a system of personalized prices $\{p_{Ni}^L\}$ can be associated such that $p_N = \Sigma p_{Ni}^L$ and $\tau_i = p_{Ni}^L x_N$, for each i. It also follows from the assumption just mentioned that $\lambda_i - \tau_i = \mu\lambda_i$, for each i. Therefore, with each public competitive equilibrium under a uniform proportional income tax rule, an allocation $(\{x_{Mi}\}, x_N)$ and a price system $(p_M, \{p_{Ni}^L\})$ can be associated such that

(a') $M[(p_M, p_{Ni}^L); \lambda_i]$ supports $C_i(x_{Mi}, x_N)$ in (x_{Mi}, x_N).

(b') $H(p)$ supports Y in $(\Sigma x_{Mi}, x_N)$.

(c') $p = (p_M, \Sigma p_{Ni}^L)$.

This allocation is a Lindahl equilibrium.

Complete separation of individual benefit from a bundle of public goods supplied and the individual cost of that bundle (given some distribution of wealth), is accepted by Fourgeaud (1969). Since his model comes close to the descriptive models usually given in textbooks about public finance, it is also a useful model for comparing various approaches and definitions.

Fourgeaud's economy consists of the usual consumption structure $(X_i, \lesssim_i, \lambda_i)$, a production multifunction $Y_{Mj}: R_+^n \to R_+^m$ for each producer and a production multifunction $Z_N: R_+^m \to R_+^n$ for the state. A complete preordering $\lesssim_g$ is defined on the range of Z_N. Prices exist only for private goods: the value of commodities supplied by the government is equal to the value of the input of private goods. This point of view is also accepted in the national accounts.

Further, the following system of redistribution and taxation rates is given in the economy: indirect taxes on the value added of private commodities, α (which may vary over commodities); direct taxes on net profits of firms, θ_{gj}, and on income, β_i; distribution of net profits of firms over consumers, θ_{ij}; and social transfers, s_i.

The assumptions are summarized by Fourgeaud in table 5.5.1.

Next, let λ_i be the i-th consumers net income to be spent on private goods and λ_g the government's income to be spent for its own consumption. Then:

$$\lambda_i = r_i - \beta_i r_i - \alpha p_M w_{Mi} \quad \text{and}$$
$$\lambda_g = r_g - \sum_i s_i .$$

It is evident that $\Sigma\lambda_i + \lambda_g = p_M(y_M + w_M)$.

Fourgeaud's conditions for equilibrium can be formulated as follows:

(a) $M(p_M; \lambda_i)$ supports $C_{Mi}(x_{Mi}, z_N)$ in x_{Mi}, for each i.

(b) For all $\bar{z}_M \in M(p_M; \lambda_g)$ it is true that $\text{Int}[Z_N(\bar{z}_M) \cap C_{Ng}(z_N)] = \phi$ and for $z_M \in M(p_M; \lambda_g)$ it is true that $z_N \in Z_N(z_M)$.

Table 5.5.1. Fourgeaud's system of national accounts

	expenditures			income		
	firms	consumers	state	firms	consumers	state
1. production				$\sum_j p_M y_{Mj}$	$\sum_i p_M w_{Mi}$	
consumption		$\sum_i p_M x_{Mi}$	$p_M z_M$			
2. redistributions:						
indirect taxes	$\alpha p_M y_M$	$\alpha p_M w_M$				$\alpha p_M (y_M + w_M)$
direct taxes	$\sum_j \theta_{gj}(1-\alpha) p_M y_{Mj}$					$\sum_j \theta_{gj}(1-\alpha) p_M y_{Mj}$
		$\sum_i \beta_i r_i$				$\sum_i \beta_i r_i$
profits of firms	$\sum_j \theta_{ij}(1-\alpha) p_M y_{Mj}$				$\sum_{ij} \theta_{ij}(1-\alpha) p_M y_{Mj}$	
social transfers			$\sum_i s_i$		$\sum_i s_i$	
	$p_M y_M$	$\sum_i r_i$	r_g	$p_M y_M$	$\sum_i r_i$	r_g

(c) $H[(1-\alpha)p_M ; (1-\alpha)(\sum_i \lambda_i + \lambda_g)]$ supports $\{\sum_j Y_{Mj}(z_N) + w_M\}$ in $y_M + w_M$.

(*d*) $y_M + w_M = z_M + \sum_i x_{Mi}$.

These conditions imply that consumers as well as government choose a maximal element, given their respective budget sets of private goods. Each firm maximizes profit according to the post-tax-prices. If the indirect taxes (α) vary over the commodities then Pareto optimality is not necessarily realized, even in the subspace of private goods. It is evident that this equilibrium need not be Pareto optimal if the public goods are also considered.

Since Fourgeaud's description of the economy is based upon the system of national accounts, an appropriate context is provided for comparing some definitions and concepts.

Firstly, it is not necessarily true that the government (the public sector) produces only public goods in the sense used in this book. This problem can be solved by considering z_M as representing the net consumption by the government.

Secondly, what is meant by public expenditures? If these are equal to the taxes received ($= r_g$), then social transfers are also included, which transfers are spent, however, on private goods. The amount spent on public goods is equal to λ_g, although in this case the fact of redistribution itself is considered a costless public good.

The same kind of problem arises in determining consumers' income. If this is said to be equal to the amount received ($= r_i$), then the taxes paid by firms are not considered to be paid on behalf of owners or consumers. If it is put equal to the amount the individual consumer can spend on private goods ($= \lambda_i$), then the consumer is completely disengaged from the finance of public goods.

The crucial question is: who has to finance the public goods level (c.q. sector)?

According to Fourgeaud's model, the public sector (r_g) is partly financed by firms, $\sum_j [\alpha + \theta_{gj}(1-\alpha)] p_M y_{Mj}$, and partly by consumers, $\sum_i (\alpha p_M w_{Mi} + \beta_i r_i)$. Since social transfers are also financed by the same sources, nothing more can be said about who finances the public goods level (λ_g). The (real life) solution is found between the extremes: either by shifting all value added to consumers and thereafter taxing or redistributing the incomes, or by shifting all value added to the government which determines individual budgets $\{\lambda_i\}$ and the public goods value $\{\lambda_g\}$, according to some agreed rules.

Since the model given in section 5.4 – the economy E_4 with a two-level price equilibrium – also allows for decentralized consumer decisions and prices on the public level, an attempt will be made to analyze the above question for the economy E_4.

Firstly, it has been noticed that enterprises have to pay duties for public goods which improve their production and efficiency. These effects have been implicitely taken into account above, but can also be explicitly treated as has been done by Kaizuka (1965) and Fabre-Sender (1969).

Further, it remains necessary for the production side of the economy that all (marginal or competitive) prices be known, in order to attain an efficient allocation in which no productive forces are wasted.

It is, however, not necessary that the income distribution be determined according to the price or productivity of the production factors owned by consumers. A complete separation of income distribution and productivity is, at least theoretically, possible, as well as a complete separation between the determination of a bundle of public goods and the financing of that bundle.

Two important constraints are present which limit the degree of separation: the problem of equity and the problem of incentives.

Both problems depend on human values, views and motives and are historically and culturally determined. If people are motivated to work only by an increase in their own share of commodities (and if some other conditions are satisfied!), then the Walrasian exchange economy can be considered as solving the incentive problem.

The equity problem can be solved only if some group of people (for whatever reason, e.g. insurance) is willing to accept a change in the income distribution without producing less. However, if some group thinks that the burden is too heavy, or is going to 'take it easy' for any other reason, then the economy starts to get in trouble.

Therefore, the income policy of a government should take both problems into consideration. And it seems that more detailed studies about what is considered fair and rewarding for various groups of consumers and producers can hardly be missed.

The considerations above are more easily realized in a distribution economy than in an exchange economy. The actual situation, however, is a mixture of both although it is so that the distribution economy gives a better description of an economy in which acquired rights are declared inviolable.

Finally, it should be noted that important contributions to the theory of taxation have recently been made by Diamond and Mirrlees (1971), Mirrlees (1971, 1972) and Baumol (1972), among others. This theory is closely connected with the theory of second best, as has been pointed out by Hahn (1972).

6. The organization of economic decisions

6.1. ALLOCATION MECHANISMS

One of those concepts which intuitively have an obvious meaning, but are hard to define rigorously, is that of decentralization. Decentralization is a property of an organization set up to achieve a specific goal. In an economy this is done to arrive at an allocation of commodities over producers and consumers.

In real life, this problem is solved, and the organization which is used to realize an allocation is called an *economic system*. Economists have developed models of economic systems, e.g., the competitive market economy. These models can often be identified by the rules through which an allocation is attained and are therefore called mechanisms or processes for resource allocation, in short, *allocation mechanisms*. A well known example of such an allocation mechanism is the competitive market mechanism guided by the tâtonnement process, as developed by Walras (1877).

Political economists, however, did not only try to describe the economic system, they also tried to adapt or even to change it. This attitude is conspicuous in the works of the socialist writers of the nineteenth century. The Marxian revolution in economic theory was not so much directed towards improving the existing economic system and mechanisms, but to replacing it by another system based on collective ownership of the means of production. This revolution was based on analysis of the productive forces in society, which forces belong to what is called here the environment of the allocation mechanism.

The socialist viewpoint that an economic system can work better without private enterprise or competitive market prices and with central planning, evoked the famous 'socialist controversy' in the 1930's. This debate focussed the attention of economists on the properties of allocation mechanisms and on the requirements they made on the economic environment.

It was Pierson (1902) who first denied that socialism could abolish the value of exchange of productive services, even if it did away with the prices of these services. He wondered how socialism would organize distribution without determining the exchange-values of different goods in terms of prices. It was the statement of von Mises (1920), however, that socialism made rational calculation and allocation impossible, which really started the debate.

The answer of Lange (1936), specified by Lerner (1944), was that in a socialist system the formal principles of the tâtonnement process, based on marginal prices, could be applied just as well as in a competitive socialist economy. But the resulting allocation might be quite different. This solution for the socialist system had already been given by Barone (1908). He introduced a Ministry of Production which was supposed to take over the role of the market in a capitalist system, by adjusting imputed prices of productive services. He thus showed that 'prices' are not conceptually bound to the institution of a market.

Solutions for the calculation problem in a socialist system were not proposed until Taylor (1929) suggested an approximation process. He distinguished commodities in terms of primary factors and final products. For any given vector of factor-prices, product-prices are set so that they equal the cost of production. If the resulting demand does not equal supply, then factor-prices should be corrected according to the observed excess-demand. This procedure, however, is valid only under strong assumptions on the economy.

Other dimensions of the problem were indicated by Hayek (1935, 1945). His doubts about the socialist system were concerned not so much with the formal problems as with the practical problems. In his opinion, the main task of an economic organization was adaptation to an autonomously changing environment. It was simply impossible that 'the knowledge of particular circumstances of time and place' of individual consumers and producers could be communicated to a central agent.

He also wondered who should valuate and accept the risk involved in decision-making. Further, he mentioned the role of incentives in an economic organization (1935, p. 234). Since an economic optimum implies 'the best use of resources known to any of the members of society, for ends whose relative importance only these individuals know', the problem of the construction of an economic organization is primarily 'the utilization of knowledge not given to anyone in its totality' (Hayek, 1945, p. 520). He therefore strongly favoured a decentralized allocation mechanism.

Although all aspects important in the design of allocation mechanisms were mentioned in the socialist controversy, further analysis began only recently. One approach in this analysis is based on the statistical theory of decision-making and is formulated in terms of the 'theory of teams': J. Marshak (1955), J. Marschak and Radner (1972). The other approach is based upon the analysis of iterative procedures in programming and game theory and on the adjustment processes of a market mechanism. Its main virtue so far is the development of concepts for analyzing allocation mechanisms, as has been notably done by: Hurwicz (1959, 1972), T.A. Marschak (1959, 1968), Malinvaud (1967) and Camacho (1972).

This last approach, which will be followed here, tries to develop performance characteristics of proposed allocation mechanisms. It is evident (nowadays) that a given allocation mechanism works satisfactorily only in a specific class of economics. Such a class is defined by facts (assumptions) over which no agent in the economy has control and is called the *environment* of an allocation mechanism, e.g. technology, preferences.

Therefore, it should first be analyzed whether a certain allocation mechanism is suitable for the given environment, and second, which of the feasible allocation mechanisms has the best performance characteristics for that environment. When the problem is so defined, it is clear that an allocation mechanism can be considered as a (scarce) public good.

Feasibility can be determined objectively, optimality is a matter of taste. In order to make a choice possible, it is necessary to define concepts which permit comparison of the performance characteristics. Some results are given in the following section.

6.2. CENTRALIZATION AND DECENTRALIZATION

Decentralization is apparently a relative concept. It depends not only on the framework in which it is used, but also on explicitly naming the terms for which it is used, viz. with respect to decisions or to information.

The first attempt to formalize the concept of decentralization with regard to information was made by Hurwicz (1959). The model has been later improved (Hurwicz, 1972) and can be described as follows (with minor changes).

An allocation mechanism (or economic process, or planning procedure) consists of two phases. In the first phase, the economic agents exchange messages (proposals, indices, prices, bids, plans). In the second phase, these messages are translated into actions and into an allocation.

For the first phase the *language M* in which communication between agents takes place must be specified, and a *set of response rules F* relating information received to information sent. The second phase consists of a *set of outcome functions G*, determining the allocation on the basis of the information assembled.

Let the set of agents be $H := \{1, 2, \dots, h\}$ in which agent 1 is, for example, a government. Each agent has knowledge about its own environment e_i; a complete specification of the environment is given by $e := (e_1, \dots, e_h)$. Messages are emitted by each agent i in a sequence of stages t, denoted by $m_i^t \in M$. In each stage the message produced is written $m^t := (m_1^t, \dots, m_h^t)$. The iteration starts with m^0, to be followed by m^1, $m^2, \dots, m^{T-1}$, and ends with m^T.

For each agent i and stage t, a response rule $f_i^t : E \times M^t \rightarrow M$ is defined and written as

$$m_i^t := f_i^t(e; m^0, m^1, \dots, m^{t-1}), \qquad t = 1, 2, \dots T,$$
$$m_i^0 := f_i^0(e).$$

Similarly an outcome function $g_i : M \rightarrow A$ is defined, where A is the set of feasible allocations,

$$a := g(m^T) \text{ and}$$
$$a_i := g_i(m^T) = \text{Proj}_i\, g(m^T).$$

The response rules can make use of accumulated information. According to the theory of communication (Shannon and Weaver, 1949), all information about the given environment can then in principle be transmitted, if only T is large enough. Most models, however, have response rules of the *first-order type*, i.e. $m_i^t = f_i^t(e; m^{t-1})$.

Another simplification is obtained if the response rules are constant over time, or *fixed*, i.e. $f_i^t = f_i^1$, for $t \geqq 1$. The Walrasian allocation mechanism for example, does possess the simplifying properties just mentioned.

Comparison with the conceptual framework used by Malinvaud (1967) gives the following results. In his allocation mechanism (called planning procedure) the language consists of two symbols: $m_1 \in M_1$, the set of indicators emitted by the central agent, and $m_j \in M_2$, the set of proposals sent by the individual agents j. The response rules are fixed and cyclical, as the procedure is a dialogue between center and individuals, i.e.:

$f_1^t = f_1^2$, for t even (called adjustment rule)
$f_j^t = f_j^1$, for t odd (called response rule).

The start of the procedure is given by f_1^0 or by f_j^0. The end of the procedure is determined by f^T, by which the plan or allocation can be determined.

An *allocation mechanism* or *decision procedure* is thus specified by the set $\{H, M_i, f_i, g_i\}$ indicating for each agent $i \in H$, the language M_i, the response rules f_i and the outcome rule g_i.

Hurwicz (1959) calls a (first-order) allocation mechanism *informationally decentralized* if it has the following characteristics:

(a) *privacy*: no unit has direct information about the internal structure of the other units, i.e.,if e and $\bar{e}$ are such that $e_i = \bar{e}_i$, then $f_i(e; m) = f_i(\bar{e}; m)$.

(b) *self-relevance*: each unit specifies only the possible effects on itself, i.e. if $(e; m^t)$, resp. $(e, \bar{m}^t)$ result in m^T, resp. $\bar{m}^T$, with $g_i(m^T) = g_i(\bar{m}^T)$, then $f_i^{t+1}(e; m^t) = f_i^{t+1}(e; \bar{m}^t)$.

(c) *anonymity*: each unit is concerned only with the aggregate effects of other units on itself, i.e. if m^t and $\bar{m}^t$ are such that the messages of any two agents other than i are permuted, then $f_i^{t+1}(e; m^t) = f_i^{t+1}(e; \bar{m}^t)$.

In his 1972–paper, however, Hurwicz calls an adjustment or allocation process *informationally centralized* if, at some stage of the process, at least one of the participants comes into possession (inferred or granted) of all relevant information concerning everyone's environment and everyone's prospective actions. This definition is also given by Pikkemaat (1969). A process is *informationally non-centralized* if such a concentration of information in one agent's hands cannot occur.

Using Hurwicz' model, the following definitions are given here to formalize the concept of decentralization in relation to information and decisions.

Assume that the message $\bar{m}^T$ contains at least enough information for every agent i so that his response function gives the same value as if the agent were in possession of the environment of all agents. This does not imply that the agents know e through $\bar{m}^T$, nor that the agents need to know e to make a correct decision. It only indicates that enough relevant information is transmitted.

An allocation mechanism is called *informationally feasible* if it generates a final message $\bar{m}^T$ with the property that for each $i \in H$, $f_i^T(e_i; \bar{m}^T) = f_i^T(e)$. It follows that an informationally centralized procedure as feasible by definition.

An allocation mechanism $\{H, M_i, f_i, g_i\}$ is called *informationally efficient with regard to language M*, if it is informationally feasible and if the sequence of messages $\{m^t\}$ is such that any other allocation mechanism $\{H, M_i, \bar{f}_i, g_i\}$ with $\bar{m}_i^t \subseteq m_i^t$, and $\bar{m}_i^t \subset m_i^t$ for at least one agent i, is not feasible.

This definition requires that, given a language and a description of what information is relevant, no more information is transmitted than is strictly necessary. Which language is preferred and what information is relevant depends on other aspects, such as the complexity of the language and the privacy or anonymity it guarantees, the communication-costs and the time needed. In principle it is possible to define a preorder on the set of languages, such that an informationally optimal allocation mechanism can be chosen.

Comparison between two informationally feasible allocation mechanisms is possible if information can be translated from one language into another, and vice versa. The allocation mechanism $\{H, M_i, f_i, g_i\}$ is said to be *informationally more decentralized* than $\{H, \bar{M}_i, \bar{f}_i, \bar{g}_i\}$ if both are feasible and if the messages $\{m^t\}$ can be deduced from the messages $\{\bar{m}^T\}$, and the converse is not true.

It follows immediately that an informationally inefficient (but feasible) allocation mechanism is informationally more centralized than an informationally efficient procedure. Presumably, if an allocation mechanism exists with the three characteristics given by Hurwicz (privacy, self-relevance and anonymity), then any other comparable procedure not having these properties is informationally more centralized. This conjecture is based on the fact that 'personal' information can always be made 'impersonal', but not vice versa.

An allocation mechanism is said to be *informationally decentralized* if all more decentralized mechanisms known are not feasible.

The following concept of decentralization is related to decisions or authority. Let A_i be the set of imaginable actions or decisions open for agent i in the economy. The multifunction $D_i: M^T \rightarrow A_i$ indicates the *choice set* $D_i(m^T)$ of agent i for every m^T, i.e. after all information has been transmitted. The multifunction $\bar{d}_i$ is called the *delegation multifunction*: if the choice set consists of one element, then the delegation is reduced to an order.

The *decision* or choice of agent i is determined by the decision or outcome function $d_i: E_i \times M^T \rightarrow A_i$, indicated by $d_i(e_i; m^T)$. It is also required, of course, that $d_i(e_i; m^T) \in D_i(m^T)$, for all $i \in H$ and $m^T \in M^T$. The allocation mechanism is *operationally feasible* if the decisions $(d_1, \ldots, d_h)$ correspond with a feasible allocation in the economy.

An allocation mechanism or decision procedure is now specified by

$$P := \{H, M_i, f_i, D_i, d_i\}.$$

In order to define a criterion for decentralization, the structure of delegation is described as follows. Let m^T be the final message in a decision procedure P. The set of agents which have a choice set strictly contained in the choice set of agent j is called the set of *subordinates*, i.e. $S(j) := \{i \in H \mid D_i(m^T) \subset D_j(m^T)\}$. The set of agents which have a choice set strictly containing the choice set of agent j is called the set of *governors*, i.e. $S^{-1}(j) := \{i \in H \mid D_i(m^T) \supset D_j(m^T)\}$. The agents can be individuals or bodies consisting of more than one individual.

An allocation mechanism P is *operationally more decentralized* than $\bar{P}$ if both are feasible and if the choice sets for all agents in P contain the choice sets for the same agents in $\bar{P}$, i.e. if $D_i(m^T) \supset D_i(m^T)$, for all $i \in H$. An allocation mechanism is said to be *operationally decentralized* if all more decentralized procedures are not feasible.

An allocation mechanism P is *operationally centralized* if there exists an agent, say 1, to whom all other agents are subordinated and whose choice set contains the choice sets of all other agents, i.e. $S^{-1}(1) = \phi$ and $D_1(m^T) \supseteq \bigcup_{i \neq 1} D_i(m^T)$.

Two extremes can be defined. An allocation mechanism is *totally centralized* if it is operationally centralized and the choice sets of all agents $i \in H$, not being the central agent, have only one element, i.e. $D_i(m^T)$ is a singleton for each $i \neq 1$. The allocation mechanism is *totally decentralized* if it is operationally feasible and if no agent has a governor, i.e. $S^{-1}(i) = \phi$, for all $i \in H$.

Combining the two decentralization concepts, it can be said that an allocation mechanism is *decentralized* if it is both informationally and operationally feasible and if a more decentralized procedure (for either aspect) is not feasible.

The feasibility condition in the definition implies that cases can be found which are operationally centralized, but still decentralized. For example, if only public goods are present in the economy, then a central agent is required to take a decision, but this decision can be based on an informationally decentralized decision procedure.

Informational feasibility (inter alia, also costs, time, etc.) limits centralization of an allocation mechanism and operational feasibility puts a limit to decentralization, given the same environment. Therefore, if the environment is such that no great transmission of information is required, then a more centralized procedure seems to be most efficient and appropriate.

Closely related with an informationally decentralized procedure is the problem of *true revelation* of information. If, for example, agents are taxed on the basis of the information about their own preferences, than it is hard to believe that they will reveal correct information (as has also been pointed out by Samuelson, 1954). This is the reason why the Lindahl equilibrium concept has in practice a very limited significance.

A similar problem related with operationally decentralized procedures is the problem of *implementation* of an optimal allocation. Both problems are connected with the question whether or not it is in an agent's interest to behave according to the rules of a given allocation mechanism, this is generally called the *incentive problem.* An allocation mechanism is *incentive-compatible* if the assumed rationality of agents sustains the behavioral rules of the mechanism, viz. the rules about revelation of preferences and about implementation. Incentive-compatibility is an important feature of an allocation mechanism and is very hard to establish. Even a perfect competitive market mechanism is not incentive-compatible (see Hurwicz, 1973, p. 24).

The market mechanism is a well-known example of a decentralized decision procedure in a classical environment. Consider the economy E_1 (in section 5.1) and the decision procedure $P = \{H, M, f_i, D_i, d_i\}$. The language M consists of bundles in R^n and price-vectors in R^{n*}. The response functions are demand functions $f_i : R^{n*} \to R^n$ for demanders and price-adaptation functions $g_i : R^n \to R^{n*}$ for suppliers, correcting prices proportionally to excess demand. The outcome functions are the demand functions when equilibrium prices are attained. This procedure P has been called a *tâtonnement process* by Walras (1874), because the market scans various prices until the equilibrium prices are found.

Property 6.2.1.

Let the competitive market mechanism in the economy E_1 be defined by a tâtonnement process and assume that this process is convergent. Then this process is informationally feasible and non-centralized and operationally totally-decentralized.

Proof

Informational feasibility follows from the fact that at the final message, with equilibrium prices all agents are able to choose a bundle which belongs to a Pareto optimal allocation, and no additional information can change this allocation without harming some agent. Since no central agent exists, the process is informationally non-centralized. The process is operationally decentralized as any enlargement of a choice set (budget set) contradicts an equilibrium condition. Total decentralization follows from the absence of a central agent. □

The convergence assumption for a tâtonnement process is by no means superfluous. Several authors have tried to establish convergence for an acceptable class of environments. The studies of Arrow and Hurwicz (1960) and Uzawa (1961) have resulted in sufficient conditions for convergence of a certain kind, but these conditions are rather restrictive.

It should be noted that the market mechanism implies that all information about other agents (their preferences, technologies and resources) is transmitted through prices and excess demand. It is assumed, however, that these data are costless and equally available to all agents. This assumption is acceptable only for simple and stable markets.

In the following two sections an outline is given of some characteristic procedures for classical and non-classical environments (see also Ruys, 1971b).

6.3. PROCEDURES WITH A SOCIAL PREFERENCE ORDERING

The allocation mechanisms which have been proposed as alternatives to the competitive market mechanism are almost always more centralized than the market mechanism. The preference of socialist authors for central planning was motivated mainly by their conception of consumption and of the social ownership of production factors. But particularly non-socialist authors designed planning procedures: firstly for the classical (convex) environments and later for non-classical environments with external effects and increasing returns.

The allocation mechanisms treated in this section are all operationally centralized, i.e. there exists a central agent or a central board which ultimately determines the allocation. For this reason allocation mechanisms are also called *planning procedures*.

Informationally, however, these procedures are decentralized. Until recently, most of the procedures developed assumed central knowledge of consumer preferences. In these cases, social preferences are given, or represented by a social objective function, and the information about the technologies of producers is dispersed over all non-central agents. Starting with these kinds of procedures, the following three approaches can be distinguished with regard to the language used by agents:

(a) the central agent transmits data (indices) in terms of prices, and the other agents respond with quantitative propositions.

(b) the central agent gives quantitative indices to other agents and receives price-propositions from those agents.
(c) both variables are sent and received.

The first approach was suggested by economists who realized that a central agent can formally replace a market and simulate all operations that take place in markets. The most specific suggestions were given by Lange-Lerner (see section 6.1) and analyzed by Arrow and Hurwicz (1960). In their case the central agent revises prices according to the tâtonnement process, which is mathematically represented by a gradient method. The principle of tâtonnement has some serious drawbacks, as has been demonstrated by Negishi (1962) and Malinvaud (1967). Firstly, for convergence quite restrictive assumptions are needed, but even then, the utility derived at each subsequent stage is not necessarily monotone increasing over time. Further, there is no guarantee that when the procedure is stopped at a finite stage, the excess demands are zero and the plan is feasible. These features are intrinsically related with the tâtonnement procedure.

Therefore, other methods have been analyzed which have better performance characteristics. Dantzig and Wolfe (1960, 1961) have developed an iterative procedure of information exchange between center and other agents, in which decisions are based on linear programming. If, in a convex environment, the center applies the programming method rather than the tâtonnement method, then the allocation obtained at each iteration is feasible and the procedure is monotone in the sense given above. This approach has been extended by Baumol and Fabian (1964) and by Malinvaud (1967).

The generalization given by Malinvaud is based on accumulation of information. At every step the center learns more about the production set of each agent. Messages sent by the center are determined by response rules which are no longer first-order rules (see section 6.2), and thus deviate from the usual planning or allocation procedures. Since the center is gradually building up a picture of each production set and must solve at each stage a mathematical programme, an enormous computer capacity is needed.

The difficulties involved in defining the concept of decentralization are clearly shown for this procedure. Since the center is gradually informed about the individual environment of all agents (their production sets), and since the decision about the allocation or plan is made by the center, Malinvaud's procedure is both informationally and operationally centralized. On the other hand, it meets the characteristics required by

Hurwicz for an informationally decentralized procedure, and it is also called decentralized by Malinvaud. This is formally in accordance with the definitions given in section 6.2 if the procedure is cut off before *all* relevant information is transmitted to the center, as will be the case in practice and as is implicitly assumed by Malinvaud. However, Malinvaud has used the term 'decentralized' to indicate initial dispersion of information.

The second approach – see (b) above – turns on sending quantity messages from the center to the agents, and price messages from agents to the center. Some authors (as Hurwicz, 1973, p. 7) contend that this approach appears to be more in line with many observed planning practices. The planning practice in a socialist economy has inspired Kornai and Lipták (1965, p. 143) to develop an allocation mechanism called 'two-level planning', in which the center assigns allocations and the agents (sectors) compute and transmit shadow prices.

The model of *Kornai and Lipták* can be summarized as follows. Let the economy consist of f sectors, each having a polyhedral production set depending on the allocations y_i assigned by the center:

$$X_i(y_i) := \{x_i \in R^n | A_i x_i \leqq y_i\}.$$

Total allocations of resources and directives to the sectors are equal to $\bar{y}_0 := \Sigma y_i \in R^m$. The objective function consists of maximizing the value of activities $x = [x_1, \ldots, x_f] \in R^{fn}$ at prices $p = [p_1, \ldots, p_f] \in R^{fn*}$, which prices are foreign exchange returns of the corresponding activities. Let $A := [A_1, \ldots, A_f]$ be the activity matrix of the economy, and $y := [y_1, \ldots, y_f]$ the allocation of y over the sectors.

The objective of a sector is to maximize $p_i x_i$ over $X_i(y_i)$ and the objective of the center is to allocate $\bar{y}_0$ such that the sum of sector profits is maximized. If the optimal allocation $\tilde{y}$ is found, then

$\sum_i \max p_i x_i$, over $x_i \in X_i(\tilde{y}_i)$, $=$ $\max px$, over $x \in X$, where $X = \{(x_1, \ldots, x_f) | Ax \leqq \bar{y}_0\}$. This implies that the center can solve the activity problem of the whole economy through an optimal allocation of resources and directives, $\tilde{y}$.

The optimal allocation of resources and directives is derived via the following planning procedure.

The dual problem for each sector is: $\min q_i y_i$, over $q_i \in Q_i$, where $Q_i := \{q_i \in R^{m*} | q_i A_i \geqq p_i\}$. Let $Q := [Q_1, \ldots, Q_f] \subseteq R^{fm*}$ and $0^+ Q$ be the recession cone of Q. Then the set of resource allocations from which the center must choose is defined by $Y := \{(y_1, \ldots, y_f) \mid y \in (0^+ Q)^*_+$ and $\Sigma y_i = \bar{y}_0\}$.

Now, at each stage t, the center sends message $\bar{y}^t$ and the sectors the message $\bar{q}^t$, calculated as follows. The center determines $\tilde{y}^t$ such that $\tilde{y}^t$ maximizes $\bar{q}^{t-1} y$ over $y \in Y$ and $\bar{y}^t$ such that $\bar{y}^t = [(t-1)/t]\bar{y}^{t-1} + (1/t)\tilde{y}^t$. The sectors determine $\tilde{q}^t$ such that $\tilde{q}^t$ minimizes $q\bar{y}^t$ over $q \in Q$ and $\bar{q}^t$ such that $\bar{q}^t = [(t-1)/t]\bar{q}^{t-1} + (1/t)\tilde{q}^t$.

This strategy defines a fictitious play method for a polyhedral game (Y, Q) with solution

$$\tilde{q}\tilde{y} = \max_{y \in Y} \min_{q \in Q} qy = \max_{x \in X} px = p\tilde{x}.$$

Kornai and Lipták have shown that the procedure is convergent (although not finitely) to the optimal allocation of resources and directives, $\tilde{y}$. The prices $\tilde{q}$ are such that $\tilde{q}_i = \tilde{q}_0$ for each sector i, and $\tilde{q}_0$ minimizes $q_0 y_0$ over $\{q_0 \mid q_0 A \geqq p\}$.

They also contend that the computational requirements are much easier to meet than those involved in computing the central programmes in the procedure outlined by Dantzig and Wolfe. The speed of convergence, however, is rather slow. True revelation is guaranteed if the sectors have to pay the prices which they report to the center. The procedure is informationally and operationally non-centralized, since the center does not need to know the A_i to decide about x_i. However, the center is supposed to known Y and thus $0^+ Q_i$, the sets of resources indispensable for each sector.

Another example in the second approach is proposed by Weitzman (1970), who has designed an allocation procedure dual to the one given by Malinvaud (1967), mentioned above. The center builds up a picture of the production sets through accumulation of information received from shadow prices. The procedure has the same properties as Malinvaud's procedure, except that feasibility is only attained in the limit.

In the class of economies in which only information about production is dispersed, the third category of allocation procedures – see (c) above – contains messages with both price and quantity information.

The extra information can be used in a classical, convex environment to improve performance characteristics. E.g., Younes (1972) lets the center indicate prices and an output neighborhood, in which costs are minimized. Presumably, the procedure is then faster and easier to compute for the agents.

But some mixed procedures are also appropriate for non-classical environments. Heal (1969, 1971) considers an economy with increasing returns. The center assigns input-quantities to each firm and determines output-prices. The firms report output-quantities and marginal productivities of inputs. The center adjusts inputs according to the 'excess productivity' of firms and equals the output-prices to the marginal social valuations. The firms respond by raising (not maximizing) output-value via substitution of outputs.

Because the response rules are continuous in time, the procedure is monotone convergent, although the center adapts a gradient method. In addition, all functions have finite partial derivatives and are differentiable.

Aoki (1970) has proposed a procedure for an economy with public goods. The center assigns the quantities of public goods to be supplied and the prices of private goods. The firms transmit marginal productivities of public goods and net demands for private goods. The response rules are based on a gradient method for the center and programming methods for the firms. Under rather strong assumptions, the procedure is monotone convergent, exactly as the Arrow-Hurwicz process mentioned above.

Procedures handling an economy with public goods and individual preference orderings are treated in the next section.

6.4. PROCEDURES WITH INDIVIDUAL PREFERENCE ORDERINGS

The problem with a social preference ordering is its construction. Unless one delegates this to a single central planner, one is always confronted with Arrow's famous 'impossibility theorem'. Arrow (1951) has shown that no rule or procedure exists for deriving a social preference ordering from a given set of individual preference orderings, given some weak conditions. Therefore, it seems worthwhile to design and analyse allocation mechanisms which directly relate individual preferences and the (final) allocation in the economy.

Just as in the previous section, three kinds of procedures will be distinguished: one in which the center uses prices to inform the other agents, one in which quantities are sent by the center, and one in which the center's indicators are both prices and quantities. All models given here include public goods.

A procedure suggested by *Lindahl* (1919) for attaining an equilibrium in an economy with public goods belongs to the first kind of approach.

Lindahl's solution has been specified by Johansen (1963) and by Malinvaud (1971), whose model will be given here.

Let the economy consist of 2 agents and 2 commodities: the public good, which quantity will be denoted by z and the private good, consumed in quantities x_1 and x_2 by the two agents respectively. An allocation is said to be feasible if z, x_1 and x_2 are non-negative and if the sum of these quantities is equal to a given number w.

Each agent i has a preference ordering $\lesssim_i$ on the commodity space R^2_+, which is extended to the private goods extension R^3_+ in fig. 6.4.1. Let each agent i have an income λ_i and let the prices of the commodities be equal to 1. The price of the public good is to be paid by both agents together, each share being p_i.

The 'demand' for public goods can then be expressed as a function of prices and the intersection of the two opposedly drawn demand curves (see fig. 5.2.1) defines the equilibrium prices $\tilde{p}_1$, $\tilde{p}_2$ and the equilibrium supply of the public good $\tilde{z} = \tilde{z}_1 = \tilde{z}_2$.

Such an equilibrium can be represented in a three dimensional commodity space (see point C in fig. 6.4.1). In this space, the two indifference curves have a cylindrical shape orthogonal to each other: the continuous lines indicate the intersection of these curves with the set of feasible allocations.

Fig. 6.4.1. Lindahl equilibrium with one private and one public good: $C = (\tilde{x}_1, \tilde{x}_2, \tilde{z})$

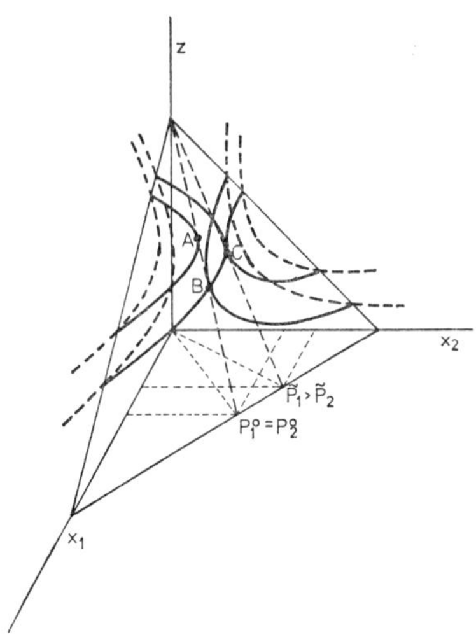

The budget constraints for each consumer are:

$$x_1+p_1 z=\lambda_1 \quad \text{and} \quad x_2+p_2 z=\lambda_2, \quad \text{with}$$
$$p_1+p_2=1 \quad \text{and} \quad x_1+x_2+z=w.$$

In this model, the following procedure is defined.

Language
The center indicates individual prices to be paid for public goods, p_i^t, and the agents propose quantities of public goods, z_i^t.

Start
The center determines an income distribution λ_i and initial prices p_i^0 (in fig. 6.4.1, $p_1^0=p_2^0=0.5$).

Iterative response rules
The agents determine their optimal consumption bundle (x_i^t, z_i^t) and propose z_i^t to the center (in fig. 6.4.1, allocation *A* is proposed by 1, and *B* by 2).

When the two proposals do not coincide, the price is raised for the agent requesting the higher quantity and correspondingly lowered for the other agent, i.e. when $z_1^t \geqq z_2^t$, then $p_1^{t+1}=p_1^t+\alpha(z_1^t-z_2^t)$ and $p_2^{t+1}=p_2^t-\alpha(z_1^t-z_2^t)$.

End
The center determines the quantity of public goods if both proposals coincide, i.e. $\tilde{z}=z_1^T=z_2^T$ (allocation *C* in fig. 6.4.1).

It has been noted by Malinvaud (1971, p. 102) that this procedure is unfair to the agent who most needs public consumption, because the other agent's demand is taken as the starting point in the adjustment rule. As can be seen from fig. 6.4.1, the process is supposed to start at point *B* instead of point *A*, both being feasible with the initial income distribution and prices. Here after, price corrections lead to – for example – point *C*. If the path from *B* to *C* is continuous, then the utility received by agent 1 will not decrease further; but it is agent 2 who benefits from the procedure.

The procedure is also not satisfactory regarding the correct revelation of preferences, since both agents will benefit individually from underreporting their demand.

A completely different approach has been suggested by *Bowen* (1943). He also defines an optimal supply of public goods by the condition that the sum of the marginal rates of substitution equals the marginal cost of a given public good. To solve the problem of measuring marginal rates of substitution, he puts forward a procedure (called 'voting') under the following assumptions:

1. All agents actually vote and express a preference which is appropriate to each agent's individual interests.
2. The cost of a public good is known and constant for each unit supplied.
3. The cost of a public good will be divided among all agents according to a given distribution of taxes which may depend on income e.a., but not on the quantity and composition of public goods supplied.
4. The curves of individual marginal substitution are distributed according to the normal distribution.

The procedure runs as follows:

Language

The center indicates individual tax rates, t_i, and the unit cost of each public good, p_N. The agents report quantities of public goods, x_{Ni}.

Start

The consumers determine their optimal consumption bundle, given the budget constraint: $p_M x_{Mi} + t_i p_N x_{Ni} \leqq \lambda_i$. The vector x_{Ni} is reported to the center.

End

For each public good $k \in N$, the center determines the modal (or average) demand, which quantity is called a Bowen equilibrium supply.

A Bowen equilibrium is not necessarily Pareto optimal, but its costs of information gathering are evidently much lower than the costs of other procedures proposed in this section. Bowen's approach has been empirically investigated by Bergstrom and Goodman (1973). They introduce the following extra assumption:

5. In each economy the median of the quantities demanded is equal to the quantity demanded by the citizen with the median income for that economy.

Bergstrom and Goodman have analyzed expenditures on several municipal public goods (general expenditures, police expenditures, and parks and recreation) for 826 municipalities in the U.S.A. with populations between 10,000 and 150,000 in 1960. Multiple regression was used to fit a function of the form:

$$\log E = c + \alpha \log n + \delta \log \hat{t} + \varepsilon \log \hat{Y} + \sum \beta_i X_i.$$

The symbols are defined as follows:

E : Expenditures of a municipality on a specified category of municipal service.

n : The number of households in a municipality.

$\hat{t}$: The tax share of the citizen with the median income.

$\hat{Y}$: Median income in a municipality.

X_i : Descriptive social and economic variables for a municipality.

According to Bergstrom and Goodman, under assumptions 1–5, a municipality's expenditure can be used as an observation of the quantity demanded by one of its citizens whose income is $\hat{Y}$ and whose tax share is $\hat{t}$.

The model of *Drèze and de la Vallée Poussin* (1971) belongs to the second kind of procedures, i.e. to those with quantities as indicators.

The economy $E=\{H;\,(X_i,\,u_i,\,\lambda_i);\,f(x)\geqq 0\}$ contains one private good and n public goods: $x=(x_M,\,x_N)\in R^{1+n}$. The utility functions, resp. production functions, are strictly concave – resp. convex – and continuously differentiable at least three times, and have non-negative partial derivatives u'_{ik}, resp. f'_k, for $i\in H$ and $k=0,\,1,\,\ldots,\,n$.

Define the (individual) prices as follows:

$q_0 := 1$, the price of the private good.

$q_k(x) := f'_k/f'_0$, for $k=1,\,\ldots,\,n$. $\quad q_N(x) := [q_1(x),\,\ldots,\,q_n(x)]$.

$p_{ki}(x) := u'_{ik}/u'_{i0}$, for $k=1,\,\ldots,\,n$. $\quad p_{Ni}(x)=[p_{1i}(x),\,\ldots,\,p_{ni}(x)]$.

Further, it is also assumed that $X_i=R_+^{1+n}$ and $p_{ki}(0,\,x_N)=0$, for all i, k and $x_N\in R_+^n$. The procedure can now be defined as follows:

Language

The center determines the allocation $(x_{0i},\,x_N,\,x_0)$, consisting of a distribution of the private good over the consumers, the supply of public goods and the total supply of the private good. The consumers announce personalized prices for public goods $p_{Ni}(x)$, and the producer announces the costs of public goods $q_N(x)$.

Start

The initial allocation $(x_{0i}^0,\,x_N^0,\,x_0^0)$ at $t=0$ is assumed to be a feasible programme satisfying $f(x^0)=0$.

Response rules

The agents determine the marginal rates of substitution at x^t and the corresponding prices $p_{Ni}(x^t) =: p_{Ni}^t$ and $q_N(x^t) =: q_N^t$, for $t\geqq 0$.

The center adjusts the allocation to conform the rules:

$\dot{x}_N \;=\; \sum p_{Ni}^t - q_N^t, \quad$ such that $\dot{x}_k \geqq 0$, for all k.

$$\dot{x}_0 = -q_N^t \dot{x}_N.$$
$$\dot{x}_{0i} = -p_{Ni}^t \dot{x}_N + \lambda_i (\sum p_{Ni}^t - q_N^t)\, \dot{x}_N.$$

End

If $\sum p_{Ni}^T = \sum p_{Ni}(x^T) = q_N^T = q_N(x^T)$, then the allocation x^T is (Pareto) optimal and is also a Lindahl equilibrium (cfr. the remarks in section 5.2).

The adjustment process consists of a tâtonnement (or gradient) process for commodities and a redistribution process of private goods, $\dot{x}_{0i}$. This redistribution consists of two terms: the first term compensates the consumer for the adjustment of public goods and the second term divides the surplus received from the information exchange. The procedure is thus always advantageous to all consumers.

Further, since consumers are taxed according to the personal prices they announce, under-reporting of the p_{Ni} is likely. But consumers are also compensated according to this information, which might lead to over-reporting. Drèze and de la Vallée state that these strategies define a game which has a strong equilibrium property, i.e. a Nash equilibrium for group rationality, if and only if people reveal correctly.

Finally, convergence of the procedure has been shown. For the existence of a solution, reference has been made, by Drèze to a theorem by Henry (1972), who was inspired by this problem.

The procedure can be extended to m private goods, where x_M denotes the $(m-1)$ dimensional vector exclusive of the numéraire, by defining the following response rules:

$$\dot{x}_{Mi} = p_{Mi}^t - q_M^t; \quad \dot{x}_N = \sum p_{Ni}^t - q_N^t.$$
$$\dot{x}_0 = -(q_M^t, q_N^t)(\dot{x}_M, \dot{x}_N).$$
$$\dot{x}_{0i} = -(p_M^t, p_{Ni}^t)(\dot{x}_{Mi}, \dot{x}_N) + \lambda_i(\sum_i |\dot{x}_{Mi}|^2 + |\dot{x}_N|^2).$$

This is not just a minor extension, however, since private goods are treated in the same way as public goods, which undoubtedly has negative effects upon the true revelation of preferences, as has been pointed out by Milleron (1972, p. 475).

Another procedure in this category has been proposed by *Ruys* (1970) for an economy $E_2 = \{H; (X_i, \lesssim_i, \lambda_i); Y\}$ with public goods only. The procedure is also based on a tâtonnement process, guided by the 'excess benefit' function $r(x) := \sum p_i(x) - q(x)$.

Language
The center indicates quantities, x, and the other agents propose prices, p_i and q.

Start
A feasible bundle of public goods, x^0, is announced by the center.

Response rules
The agents determine the optimal prices $p_i(x^t) =: p_i^t$ and $q(x^t) =: q^t$, for each $t \geqq 0$, according to the price multifunctions defined in sections 3.7 and 4.1.

The center determines the excess benefit $r^t = \sum p_i^t - q^t$, and the bundle $x^{t+1} = \alpha(x^t + \beta r^t)$, where α is chosen such that x^{t+1} is efficient, and β is chosen such that $|r(x^{t+1})| \leqq |r(x^t)|$.

End
The equilibrium bundle is attained if $r(x^T) = 0$.

The center thus determines a new bundle of public goods only if the distance from what is considered optimal at the new stage, is smaller than the distance from what was considered optimal at the previous stage. The determination of β may therefore require some intermediate exchange of information.

Convergence of the procedure depends on the behavior of the excess benefit function. Conditions have to be imposed in order to preclude that the only bundle fulfilling $|r(x^{t+1})| \leqq |r(x^t)|$ is the bundle with $|r(x^{t+1})| = |r(x^t)| \neq 0$.

If the adjustment rule is continuous, i.e. $\dot{x} := (dx_1/dt, \ldots, dx_n/dt) = \beta r(x)$, then continuity and lower boundedness of $r(x)$ is required, together with a gross substitutability condition. A theorem of Arrow and Hurwicz (1960) then implies global stability in the system $\dot{x} = \beta r(x)$. The gross substitutability condition in the finite increment version, with quantities rather than prices in the argument, can presumably be defined by: there exists $\mu \geqq 1$ such that $\mu x^{t+1} \geqq x^t$ and $\mu x_k^{t+1} = x_k^t$ imply $r_k(x^{t+1}) \geqq r_k(x^t)$, for all public goods $k = 1, \ldots, n$.

Since consumers do not have to pay taxes according to the prices they announce, they will certainly not under-report their valuations p_{Ni}. On the other hand, over-reporting one public good implies the under-reporting of one or more other public goods in the price vector p_{Ni}, since total value $p_{Ni} x_N = \lambda_i$, and λ_i are fixed for each consumer. Therefore, if consumers have no information about the prices of other individuals, the best strategy may well be true revelation of preferences. It is assumed, of course, that the n commodities considered are public goods for all participants, and that proposals are defined, for example, in terms of 'improvement of roads' and not 'improvement of that road'.

Finally, a model by *Malinvaud* (1972) will be described in order to illustrate the third kind of decision procedure, in which both prices and quantities are indicators from the center.

The economy $E = \{H, (X_i, u_i, \lambda_i);\ F, (Y_j);\ (w_M, w_N)\}$ contains m private goods and n public goods. The consumers $i \in H$ are characterized by twice differentiable utility functions u_i defined on Int X_i. The producers $j \in F$ are characterized by the production multifunctions Y_j indicating net output y_j for each bundle of public goods x_N. This technology is assumed to be represented by the twice differentiable function f_j, meeting the constraint $f_j(y_j;\ x_N) \leqq 0$.

The following notation is adopted: $x_i := (x_{Mi}, x_N)$,
$x := (x_M, x_N) = (\sum x_{Mi}, x_N)$ and $w := (w_M, w_N)$.
Demand is equal to supply if: $x = \sum y_j + w$.

Individual prices are determined by marginal valuations and expressed in terms of the first private good, the numéraire. It is assumed that the partial derivatives with respect to the numéraire, u'_{i1} and f'_{j1}, be positive for each consumer and producer. Then:

$$p_{li}(x_i) := u'_{il}/u'_{i1}\,, \qquad \text{for} \quad l = 2, \ldots, m+n\,.$$
$$q_{lj}(y_j;\ x_N) := (\partial f_j/\partial y_l)/f'_{j1}\,, \qquad \text{for} \quad l = 2, \ldots, m+n\,.$$
$$p_{kj}(y_j;\ x_N) := -(\partial f_j/\partial x_k)/f'_{j1}\,, \qquad \text{for} \quad k = m+1, \ldots, m+n\,.$$

At each stage $t = 0, 1, \ldots$ the price vectors are defined by

$$p_i^t := (p_{Mi}^t, p_{Ni}^t) := [1, p_{2i}(x_i^t), \ldots, p_{m+n,i}(x_i^t)]\,.$$
$$q_j^t := (q_{Mj}^t, q_{Nj}^t) := [1, q_{2j}(y_j^t;\ x_N^t), \ldots, q_{m+n,j}(y_j^t, x_N^t)]\,.$$
$$p_{Nj}^t := [p_{m+1,j}(y_j^t;\ x_N^t), \ldots, p_{m+n,j}(y_j^t;\ x_N^t)]\,.$$

The procedure runs as follows:

Language

The center indicates uniform prices for all commodities q, a bundle of public goods x_N, and an income distribution for private goods μ_i.

The consumers indicate their optimal consumption of private goods x_{Mi}, and their valuations (marginal willingness to pay) for public goods

p_{Mi}. The producers announce their optimal net output y_j, and their valuations (trade-off coefficients) for public goods p_{Mj}.

Start
The center announces an equitable bundle of prices q^0, and public goods x_N^0. Producers determine y_j^0 and an income distribution μ_i^0 is chosen such that $\mu_i^0 = \lambda_i[q^0 x^0 - q_N^0 x_N^0]$.

Response rules
Consumers and producers determine optimal consumption bundles x_{Mi}^t and net outputs y_j^t. This implies that $q_M^t = p_{Mi}^t$, for all consumers, and that $q^t = q_j^t$, for all producers. Next, they determine their valuations p_{Ni}^t and p_{Nj}^t for public goods.

The center adjusts the indicators as follows (superscripts t are omitted):
$\dot{x}_N = b(\sum p_{Ni} + \sum p_{Nj} - q_N)$, with $b > 0$.
$\dot{q} = b(x - \sum y_j - w)$, except for the numéraire: $\dot{q}_1 = 0$.
$\dot{\mu}_i = [-p_{Ni}\dot{x}_N + \dot{q}_M x_{Mi}] + \lambda_i[\sum p_{Ni} + \sum p_{Nj} - q_N)\dot{x}_N - \dot{q}(x - \sum y_j - w)]$.

End
If at stage T, $\dot{x}_i^T = \dot{q}^T = \dot{\mu}_i^T = 0$, then the allocation (x_{Mi}^T, x_N^T, y_j^T) is a Pareto optimal equilibrium.

The central adjustments are based upon a tâtonnement process for prices and for the quantities of public goods, and on a redistribution process of incomes for private goods. The redistribution consists of: a compensation for changing indicators, and a distribution of social profit from diminuition of excess benefit (changing social benefits), resp. excess demand (changing social costs). The procedure is thus made equitable for all consumers.

The procedure is also feasible at each stage t, and converges locally to a Pareto optimal equilibrium.

6.5. MULTI-LEVEL ORGANIZATION

The previous two sections have given an outline of several allocation mechanisms and procedures which can be found in the literature.

In this section a procedure will be proposed for arriving at a two-level equilibrium, as defined in section 5.4. In addition, an attempt will be made to generalize this procedure for an economy with local public goods. This will make a multi-level organization of information exchange and decisions necessary.

The procedure for two levels is based on the existence in the economy of a market for private goods and a referendum mechanism for public goods. A *referendum* is an institution which receives information about individual valuations of a bundle of public goods and determines a bundle of public goods which equalizes aggregate or social benefit to social cost.

The *center* in the economy consists of two independent institutions: the market and the referendum. The procedure is described as follows.

Language

The center indicates prices for private goods p_M, quantities for public goods x_N, and the share of private goods μ in the economy.

The consumers announce their individual demands x_{Mi}, and valuations p_{Ni}. The producers propose net production vectors y_j.

Start

The center indicates a feasible x_N^0 and μ^0 such that $0<\mu^0<1$. Then $\lambda_{Mi}^0 := \mu^0 \lambda_{Mi}$ and $\lambda_{Ni}^0 := (1-\mu^0)\lambda_{Ni}$.

Response rules

I. *Equilibrium within levels*

At each stage t, the market mechanism generates an equilibrium for private goods $(p_M^t, (x_{Mi}^t), x_M^t)$, such that $H(p_M^t; \mu^0)$ supportingly separates $f(x_N^{t-1})$ and $\sum C_{Mi}(x_{Mi}^t, x_N^{t-1})$ in x_M^t.

At each stage t, the referendum mechanism generates an equilibrium for public goods $(x_N^t, (p_{Ni}^t), p_N^t)$, such that $H(x_N^t; 1-\mu^0)$ supportingly separates $(1-\mu^0) * f^{-1}(x_M^{t-1})$ and $\sum \lambda_{Ni}^0 * C_{Ni}(x_{Mi}^{t-1}, x_N^t)$ in p_N^t.

II. *Feasibility between levels*

At stage $t = 1, 3, \ldots$, the market indicates prices p_M^t, consumers determine their consumption x_{Mi}^t, and producers determine the net output of private goods and the production possibility set of public goods. At stage $t = 2, 4, \ldots$, the referendum indicates quantities x_N^t, based on the valuations p_{Ni} and p_{Nj} proposed by consumers and producers, with quantities of private goods fixed. A sequence of prices and quantities is chosen which converges: $x_N^0, p_M^1, \ldots, x_N^t, p_M^{t+1}, \ldots, x_N^T, p_M^T$.

III. *Income redistribution between levels*

The center indicates prices $p := (p_M^T, p_N^T)$ to the producers who maximize profits over their complete production sets. The producers propose $\tilde{y} \in Y$. If $\tilde{y} \neq y := (x_M^T, x_N^T)$, then $p\tilde{y} > 1$ and the center determines $q := (\alpha p_M^T, \beta p_N^T)$ such that $qy = 1$ and $q\bar{y} \leqq 1$ for all $\bar{y} \in Y$. (See also fig. 5.4.2.) This is done by an iterative procedure in which at each stage s, the center proportionally lowers all prices at which producers can make more profit, by lowering the share of income, μ^0 resp. $(1-\mu^0)$, to $\alpha^s \mu^0$ resp. $\beta^s(1-\mu^0)$, until $q^s := (\alpha^s p_M^T, \beta^s p_N^T)$ is such that $q^s \tilde{y} = q^s y = 1$. Then the center indicates q^s, the producers propose y^s, until no production vector can be proposed with $q^T y = 1$ and $q^T \tilde{y} > 1$, for some $\tilde{y} \in Y$.

End

The center indicates a bundle x_N^T of public goods, a price vector p_M^T of private goods, and a production share $\mu := \alpha\mu^0$ of private goods. The corresponding allocation $[\{(x_{Mi}^T, x_N^T)\}, y^T, \{(p_M^T, p_{Ni})\}, q^T]$ is a two-level price equilibrium.

The total procedure consists of three partial procedures. The properties of the partial processes under I have been mentioned in sections 6.3 and 6.4. Although it can be said that the process under II is similar to an actual (one stage) decision process in some economies, the process is convergent only under rather restrictive conditions. From the mathematical point of view, this is certainly the weakest part of the procedure. The incentive and communication requirements, however, are minimal.

Convergence of part III of the procedure is easily verified. It can be seen that if redistribution does not takes place, another bundle will be produced and equilibrium within the levels will be disturbed. Since redistribution via the lowering of an income share is hard to perform, it is easier to obtain the same result by increasing some prices more than others, which implies inflation.

It is evident that the procedure is not centralized (informationally or operationally). On the contrary, since a more decentralized procedure is not feasible, the procedure is decentralized.

The concept of a two-level price equilibrium and the procedure proposed above can be generalized for an economy with several local public goods (see section 4.5). Let the economy E_5 be defined by:

$$E_5 := \{H, (X_i, \precsim_i); F, (Y); G_{kj}, (\lambda_{kj}, \lambda_{ji})\},$$

in which G_{kj} denotes the set of agents in referendum j collectively using commodity k. The set G_k is the set of referendums for k, which simultaneously is the set of participants in the market for commodity k. It follows immediately that commodity k is a private good if $G_k = H$ and

$G_{kj}=\{1\}$, and that commodity k is a public good if $G_k=\{1\}$ and $G_{kj}=H$. The weight distribution within is given by λ_{ji}, i.e. $\sum\lambda_{ji}=1$ for $i\in G_{kj}$. The income distribution between referendums relative to good k is given by λ_{kj}, i.e. $\sum\lambda_{kj}=1$ for $j\in G_k$.

Assume that there exist three levels in the economy:

1. a referendum for the set N of public goods, i.e. for all $k\in N$, $G_{kj}=G_N$, for which the net product share λ_N is given: $\lambda_N=\sum\lambda_{kj}=\sum\lambda_k$, for $k\in N$.
2. g referendums for the set L of local public goods, i.e. for all $k\in L$, there exist referendums $G_{kj}=G_{Lj}$, $j=1, \ldots, g$, for which the net product share λ_{Lj} equals: $\lambda_{Lj}=\Sigma\lambda_{kj}$, $k\in L$.
3. a market for the set L of local public goods and the set M of private goods, in which all agents (referendums and individuals) can buy and sell, given the income distribution λ_{Lj} and $\lambda_{Mi}=\lambda_{Mj}=\sum\lambda_{kj}$, for $k\in M$.

A *three-level price equilibrium* is defined to be the allocation $[(x_{Mi}, x_{Lj}, x_N), y, (p_M, p_{Li}, p_{Ni}), q]$ in the three-level economy E_5, such that for each j:

(a–1) the choice sets $M(p_M; \lambda_{Mi})$ support $C_{Mi}(x_{Mi}, x_{Lj}, x_N)$ in x_{Mi}, with $i\in G_{Lj}$.

(a–2) the choice sets $M(p_L; \lambda_{Lj})$ support $C_{Li}(x_{Mi}, x_{Lj}, x_N)$ for $i\in G_{Lj}$, in x_{Lj}.

(a–3) the choice sets $M(x_{Lj}; \lambda_{Lj})$ support $C^*_{Li}(x_{Mi}, x_{Lj}, x_N)$ in p_{Li}, for each $i\in G_{Lj}$.

(a–4) the choice set $M(x_N; \lambda_N)$ supports $C^*_{Ni}(x_{Mi}, x_{Lj}, x_N)$ in p_{Ni}, for each $i\in H$.

(b) $q=(p_M, p_L, p_N)$ with $p_L=\sum\lambda_{ji}p_{Li}$, for $i\in G_{Lj}$, and $p_N=\sum\lambda_{ji}p_{Ni}$, for $i\in G_N$.
$x=(x_M, x_L, x_N)$ with $x_M=\sum x_{Mi}$, for $i\in H$, and $x_L=\sum x_{Lj}$, for $j=1, \ldots, g$.

(c) the hyperplane $H(q)$ supports Y in x.

Thus conflicts of interests between agents (individuals for (a–1) and councils for (a–2)) over quantities are brought into equilibrium on markets, and conflicts of interests between agents over valuations (see (a–3) and (a–4)) are solved in referendums. Referendums will respond on decisions made at a higher level, since the public goods supplied are available to all lower levels (see fig. 6.5.1). On the other hand, referendums will respond on 'decisions' of markets at a lower level, in order to obtain an equilibrium in which collective benefits are equal to collective costs. This interdependence is expressed in fig. 6.5.1.

Fig. 6.5.1. Uniform quantities via referendums, and uniform prices via markets in a three-level economy

level	quantities			prices		
public	x_N ↓			p_N	p_L	p_M ↑
local public ($j \in G_L$)	x_N	x_{Lj}			p_L	p_M
private ($i \in H$)	x_N	x_{Lj}	x_{Mi}			p_M

The production sector is assumed to supply all commodities at marginal prices. An enterprise, however, can be considered as being an economy itself with an internal structure as given above (see also section 7.1). The bigger an enterprise is, the more that enterprise produces public goods (if only employment), and the more it consumes public goods. It becomes an agent which is not controlled (at least not completely) by consumer markets and referendums. On the contrary, big enterprises can accumulate power by controlling consumer markets and referendums. It is clearly of interest to consumers (and also to enterpiises) to develop mechanisms which direct the enterprise to its only goal: that of producing a bundle of commodities which society considers optimal.

The two-level procedure described above can, in principle, also be applied to a three-level economy. The sequence of messages, on which basis the levels can determine their optimal programmes, becomes: x_N^t, x_L^{t+1}, p_M^{t+2}, p_L^{t+3}, x_N^{t+4}, ... If feasibility between the levels is attained, redistribution of income shares between the levels may be necessary. At an equilibrium distribution, a two level price equilibrium is achieved and all agents can choose optimal elements from their choice sets.

Again, in principle, the three-level price equilibrium can be generalized to a *multi-level price equilibrium* for an economy with more than one level of public goods. The philosophy of a multi-level organization is based on the existence of economic units regulating the supply and demand of (local) public goods. Internally, a referendum mechanism determines a cooperative equilibrium; externally, it accepts the market mechanism assigning a competitive equilibrium allocation. This implies that local public goods are marketable, i.e. comparable, replicable or exchangeable, just as private goods. Thus, the flexibility and efficiency of the original competitive economy is coupled with the stability and interdependence associated with public goods.

One of the main virtues of the multilevel organization is the distribution of choice sets (power) over many economic units. Although a national-economic mechanism (and choice set) is needed for public goods, this mechanism can be restricted to public goods only. Decentralization is possible to that level which is just high enough to include all agents affected by some economic good.

The equity problem can be separated into the problem of income distribution between consumers for private goods, the problem of income distribution between units for (local) public goods, and the problem of weight distribution within the referendums determining (local) public goods. (See also section 5.5). The third problem is rather new in economic theory, as was pointed out in section 1.2. But if a *democracy* is defined as being a system of assigning a bundle of public goods such that each participant has an equal weight in the determination of the social benefit of that bundle, then the referendum mechanism is democratic if the λ_{Ni} are uniformly distributed. Thus, a 'one man, one vote' system can be obtained.

A necessary institutional condition for this multi-level organization is, of course, the existence of markets and referendums at appropriate levels. But an appropriate attitude (or ideology) on the part of economic agents is also needed for competitive and cooperative behavior in the sense used here.

7. Extensions

7.1. BETWEEN VALUES AND RESOURCES

The environment in which public goods are defined need not of course, coincide, with a national economy. The extent of public goods produced can be larger, so that a supranational organization is required, or smaller. It also is possible that commodities are public, relative to a set of consumers only, or public, relative to a set of producers only.

An organization in which public goods of the second type are supplied is an enterprise or a concern. Commodities such as: research and development, general management, sales promotion, maintenance and repair-shop, computing and accounting services, can often be considered as public goods supplied to the various production sectors within an enterprise. The economy with public goods then coincides with the enterprise, and the 'consumers' are the various sectors which have their own preferences about the inputs of public goods, derived from their technology (see section 3.5). Which bundle is optimal for those sectors in an enterprise can be determined by a referendum mechanism, in which the weight distribution depends on each sector's contribution to gross profit.

The increasing extent and complexity of modern enterprises makes comparison between national economies and enterprises occur more often. In many cases, an enterprise can be considered as an economy, and models or concepts developed in economic theory can be applied to enterprises (see also Verheyen, 1965). If only for this reason, a certain convergence between political economy and business administration will take place, presumably.

On the other hand, although the formal structures of enterprises and national economies may become more similar, the interpretations will still show a great dissimilarity. In a national economy, the emphasis is on the consumers who want to determine a bundle of public goods on basis of their own preferences and value judgements. The consumers together are sovereign, and the central agent referred to in the referendum mechanism is not a government with its own, or even derived preferences, but purely a computing and executing institution. This interpretation fits in perfectly with the ideas developed by Rousseau (1762). The referendum mechanism constructs a 'volonté générale' of a commonwealth with universal suffrage. The central agent expresses this will of the sovereign, and executes its decisions without influencing the decisions at all.

In practice, however, the central agent consists of various institutions: e.g. a parliament determining the general will of the people, a government executing this will, and a judiciary safeguarding the rules. Each institution has its own preferences, and influences the outcome. The 'sovereign' is called to the polls with voting papers that hardly leave room to express sufficiently the voters' individual preferences about a certain bundle of public goods. This, or another, political system still works satisfactory if it is an equilibrium between the information – and time – consuming solution of Rousseau's pure democracy, and the alienating solution of an autocratic system. Such a consideration, however, belongs to political science. In this study it has been shown – from the economic point of view – that an allocation mechanism exists by which consumers' sovereignty can be realized, also when public goods are present in the economy.

The economies E described in chapters 2 and 5 are defined in the quantity or commodity space. It has been mentioned in section 1.3 and shown in section 5.1, that a dual economy or a valuation representation E^* can be associated with the economy E. This dual economy E^* is only another representation of the same economy and related with E through a formal duality operation. This duality relation connects three concepts:

$$\begin{array}{ll} \uparrow E^* & : \text{the nominator system} \\ \bullet\, E & : \text{the real system} \\ \downarrow 1 & : \text{the denominator (resources).} \end{array} \qquad (7.1.1)$$

The duality operation * can also be given a wider interpretation: viz. a process of development. The first process is compared to dialectic materialism, a process from resources and a real system into a new system: 'In a system constituting a whole there appear contradictions which prevent the system from remaining in a changeless state. The contradictions in the system induce changes leading to a readjustment which makes the contradictions disappear. But these very changes open the way to new contradictions which, in turn, induce new changes – etc. As a result, wholes can never remain in a changeless state; they must change constantly. The changes, however, show a definite direction; in other words: they represent a *process of development*. In the course of development, individual wholes combine into more complex systems, into wholes 'of higher order' which exhibit new properties and new modes of action hitherto not encountered. Thus, in the course of dialectical development *new properties* (new 'qualities') and new modes of action (new laws of behavior) come into being.' Oskar Lange (1962), p. 1–2.

According to dialectic materialism, new systems with higher qualities are generated by existing systems. The nominator system E^* in (7.1.1) can be interpreted as such a new system. It is, however, also possible to interprete E^* as a new system of valuations caused by E. For example, shortages or bottle-necks can stimulate producers to change their production technologies (Schumpeter, 1913), or stimulate consumers to change their preferences (Meadows, 1972). The process $v: E^t \rightarrow E^{*t+1}$ is called a *materialistic process*, because it departs from the system E of lower order.

The second process, $w: E^{*t} \rightarrow E^{t+1}$, from a system of higher order (the nominator system) into a system of lower order (the real system), is an inverse of a materialistic process and called a *finalistic process*. This process takes place if, for example, an existing system of valuations and capacities generates a real system using resources and possibilities present in the environment. The formulation of declarations by the United Nations, or of economic policies is based on the existence of such processes (see also Cobbenhagen, 1943, and Dalmulder, 1950). When a new system has thus been realized, it becomes a resource for a new fase in the development.

Any real system arises from the tension between an existing nominator system, and an existing denominator. An economy is thus a product of an existing system of valuations and productive forces, and the resources (or capital) left over from, or built up during previous stages.

Both processes refer to qualitatively different systems, but establish relations between elements in these systems via the same denominator (see diagram 7.1.4). These relations are called *nominal relations*. Materialistic processes induce centrifugal forces in these relations, developing new qualities. Finalistic processes embody valuations into a seizable reality, using the centripetal force of the imperfect, but mouldable denominator.

Which system, E or E^*, is of higher order, can be deduced from the analysis of the contradictions *within* each system. The contradictions can be resolved by a state of equilibrium, denoted by $\tilde{e}$ or $\tilde{e}^*$. If $\tilde{e}^*$ indicates the direction of development for the resolution of the contradictions in E, then E^* is the *system of higher order*. The state or element $\tilde{e}^*$ in E^* gives the *values* or norms of the whole considered. These values arise from countervailing forces within each system. Suppose that these countervailing forces in an economic whole are the consumption and production foıces, denoted by C and P. Then the *valuation relation* between C and P, resp. C^* and P^*, results in the state $\tilde{e}^*$:

$$\begin{array}{ll} \text{system } E^*: & C^* \longleftarrow \overset{\tilde{e}^*}{\bullet} \longrightarrow P^* \\ \text{system } E\ : & C \longleftarrow \overset{\tilde{e}^*}{\bullet} \longrightarrow P \end{array} \qquad (7.1.2)$$

In an economy, this equilibrium state or value system $\tilde{e}^*$ can be a set of prices or valuations which guide the decisions of the agents in the systems, so that their physical needs are also in a state of equilibrium.

Finally, the contradictions between nominally different forces, and between the two systems should be resolved. The relation between C^* and P, resp. P^* and C, is called a *dependence relation* and results in the state $\tilde{e}$. In the example used here, both actual consumption and actual production are dependence relations:

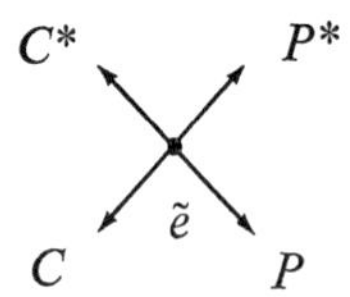

(7.1.3)

Although all elements are equivalent in the system of relations, and can be subjected to change, it is the element $\tilde{e}$ in which one is usually interested when a whole system is studied. In order to estimate in which direction change will take place, it is necessary to analyze the forces that cause the change. It is conjectured here that every force is an element in the whole system, related to all other elements. These forces generate new values and a new reality, by which – in turn – the forces are changed. The elements and the relations are composed in diagram 7.1.4, which diagram can also be found in Ruys (1963).

In this conception, an equilibrium or an outcome $\tilde{e}$ is a realization of a relation between values and resources. These three elements belong to a whole, together with the contradicting or countervailing forces which determine the equilibria. As long as each change in an element induces changes in other elements, the whole is (continuously) changing, or in a state of non-stationary motion.

Diagram 7.1.4. An economy as a changing whole

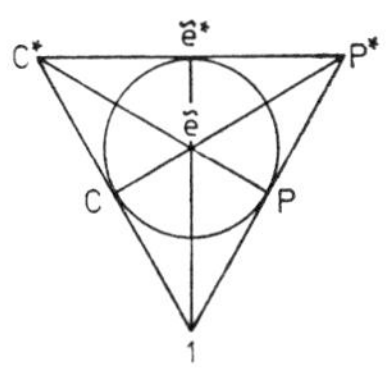

If the systems E and E^* are representations of each other (see section 5.1), then the nominal relations are in equilibrium and the whole is in a state of equilibrium or of stationary motion.

7.2. THEORY OF MOTION

Let Ω be a set of eight elements $\{x_i | i = 0, \ldots, 7\}$, and let $*$ be a function from $\Omega \times \Omega$ in Ω, whose value $*(x_i, x_j)$ is also denoted by $x_i * x_j$. The pair $(\Omega, *)$ constitutes a set of elements together with a set of relations between elements and is thus a system.

The system $(\Omega, *)$ is called a *whole* if it satisfies the following axioms:

7.2.1. The associative law holds, i.e.
$(x_i * x_j) * x_k = x_i * (x_j * x_k)$, for all $x_i, x_j, x_k \in \Omega$.

7.2.2. There exists an identity element x_0, i.e.
$x_0 * x_i = x_i * x_0 = x_i$, for all $x_i \in \Omega$.

7.2.3. Each element is its own inverse, i.e.
$x_i * x_i = x_0$, for all $x_i \in \Omega$.

7.2.4. The commutative law holds, i.e.
$x_i * x_j = x_j * x_i$, for all $x_i, x_j \in \Omega$.

7.2.5. The inverse of $*$ is a multifunction $*^{-1}: \Omega \to \Omega \times \Omega$, such that
$\bigcup_{i=0}^{7} *^{-1}(x_i) = \Omega \times \Omega$ and $\bigcap_{i=0}^{7} *^{-1}(x_i) = \phi$.

It follows that a whole $(\Omega, *)$ is an Abelian group. The relation defined in Ω is transitive, reflexive, symmetric and complete (see section 8.1), and thus an equivalence relation in Ω. Axiom 7.2.5 implies that any ordered pair belongs to the relation.

An example of a whole is the set
{0, 0, 0), (0, 0, 1), (0, 1, 0), (1, 0, 0), (1, 1, 0), (0, 1, 1), (1, 0, 1), (1, 1, 1)}, together with the operation $*$, which is defined to be addition in a number system modulo 2, e.g.
$(1, 1, 1) * (0, 0, 1) = (1, 1, 0)$. The identity element is $(0, 0, 0)$.

It is conjected here that any closed system in motion can be defined in terms of elements and relations, such that a group isomorphism between that system and $(\Omega, *)$ can be established, providing that the identity element x_0 is not defined in the system and plays a formal role in $(\Omega, *)$ only.

A representation of such a system, or a whole is given in fig. 7.2.1 (see also Sawyer, 1962, p. 201).

Fig. 7.2.1. Representation of a whole

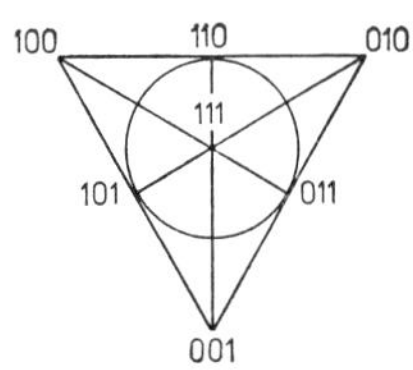

II. Mathematics

8. Basic mathematical notions and notations

8.1. SETS, RELATIONS AND MULTIFUNCTIONS

The introduction and definition of concepts in this chapter provides a list of symbols and outlines the notions and properties to which reference is made in this study.

Let A be a set of elements; the set B is called a *subset* of A, denoted by $B \subseteq A$, if each element of B belongs to A, i.e. $a \in B$ implies $a \in A$. The set B is called a *proper subset* of A, denoted by $B \subset A$, when B is a subset of A and A is not a subset of B, i.e. $B \subseteq A$ and $A \nsubseteq B$.

The *difference* between two sets A and B is defined as being the set of elements which belong to A and not to B, i.e. the set $A \backslash B := \{a \in A \mid a \notin B\}$. The *empty set* is denoted by ϕ.

Let A be a set, M a set of index numbers and $\{A_i \mid i \in M\}$ a family of subsets of A. The *union* of the sets A_i is defined as being the set of elements which belong to at least one A_i, i.e. $\bigcup_{i \in M} A_i := \{a \in A \mid \exists i \in M : a \in A_i\}$. The *intersection* of the sets A_i is said to be the set of elements which belong to all A_i's, i.e. $\bigcap_{i \in M} A_i := \{a \in A \mid \forall i \in M : a \in A_i\}$. If the index set M contains m elements, then the *cartesian product* of the sets A_i is said to be the set of all ordered m-tuples, i.e. $\prod_{i \in M} A_i := A_1 \times A_2 \times \ldots \times A_m := \{(a_1, a_2, ..., a_m) \mid \forall i \in M : a_i \in A_i\}$.

Let A be a set; the set of all subsets of A is said to be the *power set* of A, and denoted by $\pi(A)$.

Let A and B be sets. An *ordered pair* (a, b) is called an element of the cartesian product of A and B, i.e. $(a, b) \in A \times B$. *A* (binary) *relation* R is defined as being a set of ordered pairs, i.e. $R := \{(a, b) \mid a \in A \text{ and } b \in B\} \subseteq A \times B$. The set $D(R) := \{a \in A \mid \exists b \in B : (a, b) \in R\}$ is called the *effective domain* of the relation, and the set $R(R) := \{b \in B \mid \exists a \in A : (a, b) \in R\}$ is said to be the *effective range* of the relation R.

The *inverse* of the relation R is defined as being the set $R^{-1} := \{(b, a) \mid (a, b) \in R\} \subseteq B \times A$

Let $R_1 \subseteq A \times B$ and $R_2 \subseteq B \times C$ be relations; the set $R_2 R_1 := \{(a, c) \mid \exists b \in B : (a, b) \in R_1 \text{ and } (b, c) \in R_2\} \subseteq A \times C$ is then called the *composition product* of the relations R_1 and R_2.

The relation R is said to be a *relation in a set A* if $D(R) = A$ and $R(R) \subseteq A$. The following properties of a relation R in a set A are defined:
R is *transitive*, or an *ordering relation* in A, if $RR \subseteq R$;
R is *reflexive* if $\{(a, a) \mid a \in A\} \subseteq R$;
R is *symmetric* if $R = R^{-1}$;
R is *a-symmetric* if $R \cap R^{-1} = \phi$;
R is *anti-symmetric* if $R \cap R^{-1} \subseteq \{(a, a) \mid a \in A\}$;
R is *complete* if $R \cup R^{-1} = A \times A$.
Let R be a relation in A; instead of $(a, b) \in R$, we can then also write $a \lesssim b$.

If a relation R in a set A is defined, then this set is said to have the following properties:
A is *partially preordered* if R is transitive and reflexive;
A is *partially ordered* if R is transitive, reflexive and anti-symmetric;
A is *completely preordered* if R is transitive, reflexive and complete;
A is *completely ordered* if R is transitive, reflexive, anti-symmetric and complete. Examples are given below, in which R may be interpreted as a preference relation in a set S consisting of four elements $\{a, b, c, d\}$.

a, d		c, d	d, d
a, c		c, c	d, c
	b, b		
a, a			

R_1 : partial preorder

a, d	b, d	c, d	d, d
a, c		c, c	
a, b	b, b		
a, a			

R_2 : partial order

a, d	b, d	c, d	d, d
a, c	b, c	c, c	
a, b	b, b	c, b	
a, a			

R_3 : complete preorder

a, d	b, d	c, d	d, d
a, c	b, c	c, c	
a, b	b, b		
a, a			

R_4 : complete order

The set $S := \{a, b, c, d\}$ is: partially preordered if R_1 is defined in S; partially ordered if R_2 is defined in S; completely preordered if R_3 is defined in S, and completely ordered if R_4 is defined in S.

A relation R in a set A is said to be an *equivalence relation* if R is transitive, reflexive and symmetric. An *equivalence class* of a preordered set A is the subset $\{b \in A \mid (a, b) \in R \cap R^{-1}$, for some $a \in A\}$. An example of an equivalence class is the set $\{c, d\} \subseteq S$ with relation R_1, or $\{b, c\} \subseteq S$ with relation R_4. A relation R in A is called a *strict ordering relation* if it is transitive and a-symmetric.

Let R be an equivalence relation in A; instead of $(a, b) \in R$, we can also write $a \sim b$. If R is a strict ordering relation, then we can also write $a \prec b$, instead of $(a, b) \in R$.

Let A be a set *partially preordered* by the relation R. The element $a \in A$ is said to be a *maximal element* of A if $(a, b) \in R$ implies that $(a, b) \in R^{-1}$. For example, the elements b, c and d in S with relation R_1 are maximal elements of A, since neither (b, a) (b, c) and (b, d) nor (c, a), (c, b), (d, a), (d, b) belong to R_1; but both (c, d) and (d, c) belong to R_1. Let A be a preordered set and let $B \subseteq A$; the element $a \in B$ is a maximal element of B if and only if one of the following statements is true:
1. $a \lesssim b$ implies $b \sim a$;
2. $a \gtrsim b$ or $a \not\lesssim b$;
3. $a \prec b$ implies $b \notin B$.

The element $a \in A$ is an *upper bound* of $B \subseteq A$, if $b \in B$ implies that $(b, a) \in R$. A *minimal element* of a set, resp. a *lower bound*, is defined as being a maximal element, resp. an upper bound, for the inverse relation R^{-1}.

Let A be a set partially preordered by the relation R; the element $a \in A$ is said to be a *maximum* or *greatest element* of A if $b \in A$ implies that $(b, a) \in R$. The element $a \in A$ is called a *minimum* or *least element* of A if $b \in A$ implies that $(b, a) \in R^{-1}$. An example of a least element of the set $S = \{a, b, c, d\}$ above is the element a under relation R_2, R_3 or R_4.

A maximum of a set is always a maximal element of that set. If a set A is completely ordered, then a maximal element of A is also a maximum of A, which maximum is unique and denoted by $\max\{A\}$. In that case a minimum is denoted by $\min\{A\}$.

Let X and Y be sets. A *multifunction* F from a set X into a set Y is a relation defined on $X \times Y$, i.e. $F \subseteq X \times Y$. The multifunction F assigns to every $x \in D(F)$ a set in $R(F)$ defined by $f(x) := \{y \mid (x, y) \in F\}$; therefore the multifunction is also represented by the rule f and the sets X and Y, and denoted by $f: X \to Y$. This convention will be followed here. A multifunction is also called a *correspondence*, a multivalued function or a set-valued function.

Let $f: X \to Y$ be a multifunction. The *graph* of f is defined as being the set

$$G(f) := \{(x, y) \in X \times Y \mid y \in f(x)\} = F.$$

The *effective domain*, resp. *effective range*, of f is said to be the sets

$$D(f) := \{x \in X \mid f(x) \neq \phi\} = D(F);$$
$$R(f) := \{y \in Y \mid \exists x : y \in f(x)\} = f(X) = R(F).$$

The *image* of a subset $A \subseteq X$ under the multifunction f is said to be the set

$$f(A) := \{f(x) \mid x \in A\}.$$

The *inverse* of f is called the multifunction $f^{-1} : Y \to X$ defined by

$$f^{-1}(y) := \{x \mid y \in f(x)\}, \text{ for all } y \in R(f).$$

The *inverse-image* or *counter-image* of a set $B \subseteq Y$ under f is said to be the set

$$f^{-1}(B) := \{f^{-1}(y) \mid y \in B\} = \{x \mid f(x) \cap B \neq \phi\}.$$

The *upper inverse* of a set $B \subseteq Y$ under f is the set

$$f^{+}(B) := \{x \mid f(x) \subseteq B\}.$$

A multifunction $f: X \to Y$ is said to be a *function* from X into Y if $(x, y) \in G(f)$ and $(x, z) \in G(f)$ imply that $y = z$. One can associate with every multifunction $f: X \to Y$ a function $\bar{f}: X \to \pi(Y)$, by setting $\bar{f}(x) := f(x)$ for every $x \in X$. If $f: X \to Y$ is a function, then $f^{-1}(y) = \{x \mid y = f(x)\}$; the function f is said to be *univalent*, or a one-to-one function, if f^{-1} is also a function.

Let $X := X_1 \times X_2 \times \ldots \times X_n$, the cartesian product of n sets X_i; the *projection* of X on X_k is said to be the multifunction $\text{Proj}_k : X \to X_k$ defined by

$$\text{Proj}_k(X) := \{x_k \in X_k \mid (x_1, \ldots, x_k, \ldots, x_n) \in X\}.$$

It should be noted that $\text{Proj}_1[G(f)] = D(f)$ and $\text{Proj}_2[G(f)] = R(f)$.

Let P be a property of sets. A multifunction $f: X \to Y$ is said to be *point-P* if $f(x)$ has the property P for each $x \in X$.

The logic of propositions can be represented and visualized by set-theoretical symbols, applied on the following Boolean algebra of sets; see also fig. 8.1.1. Let A be the set of all elements considered; the set $P \subseteq A$ consists of those elements for which the statement p is true (P is a circle), and $Q \subseteq A$ consists of those elements for which the statement q is true (Q is a triangle). The propositions 'p and q', 'p or q', '$p \Rightarrow q$' and '$p \Leftrightarrow q$' are then true in the shaded subsets of A:

Fig. 8.1.1. Subsets of true and false propositions

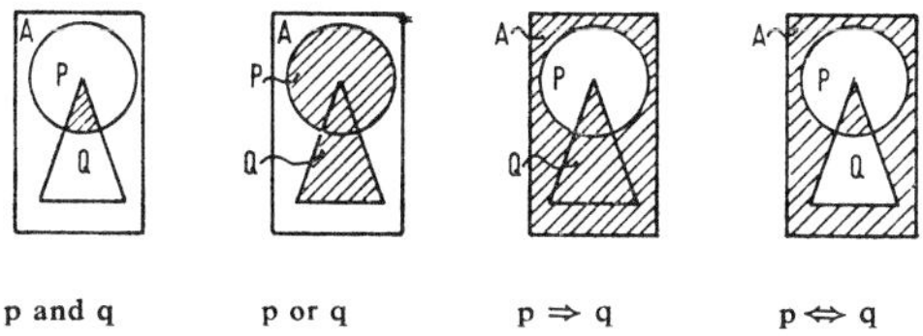

It follows, for example, that 'p and q' is always true if and only if $P \cap Q = A$; and that '$p \Rightarrow q$' is never false if and only if $P \subseteq Q$.

It can also be seen from the above figures that, 'not (p or q)' is equivalent to '(not p) and (not q)'; 'not (p and q)' is equivalent to '(not p) or (not q)'; '$p \Rightarrow q$' is equivalent to '(not p) or q'; etc.

8.2. CONTINUITY OF MULTIFUNCTIONS

Let R be the *set of real numbers*, and let $R^n := R_1 \times R_2 \times \ldots \times R_n$, the cartesian product of n sets R. The following operations are defined on elements in R^n, which are denoted by $x := (x_1, x_2, \ldots, x_n) := [x_i]$:

addition : $x+y := [x_i + y_i]$;
scalar multiplication: $\alpha x := [\alpha x_i]$, for some $\alpha \in R$.

The set R^n, which is closed under the operations of addition and scalar multiplication, is called a *real linear vector space*.

Let $f: R^m \to R^n$ be a multifunction; f is said to be a *linear transformation* if $f(x+y) = f(x) + f(y)$ and $f(\alpha x) = \alpha f(x)$.

The set R^n can also be endowed with a topological structure by defining a metric on R^n. The following operation is therefore defined on elements in R^n: inner product: $\langle x, y \rangle := \sum_{i=1}^{n} x_i y_i$.

The real linear vector space R^n with this standard Euclidean inner product is called the *real Euclidean n-space*.

The Euclidean norm or length of a vector is defined by

$$|x| := \sqrt{\langle x, x \rangle}.$$

This norm induces a metric in R^n, which is called the *Euclidean distance* function:

$$d(x, y) := |x-y| = \sqrt{\sum_i [x_i - y_i]^2}.$$

The symbol R^n will be used to indicate the metric space consisting of the

real Euclidean n-space and the Euclidean distance function. All sets in this section are subsets of R^n.
Let $\varepsilon > 0$ be a scalar and $x \in R^n$; the set

$$B(x;\,\varepsilon) := \{y \in R^n \mid d(x, y) \leqq \varepsilon\}$$

is said to be an *ε-neighborhood* of a point x, or a *ball* with center x and radius ε. Let $X \subseteq R^n$; then the set

$$B(X;\,\varepsilon) := \bigcup_{x} \{B(x;\,\varepsilon) \mid x \in X\}$$

is called an ε-neighborhood of X.

A point $x \in X$ is called an *interior point* of X, if a neighborhood $B(x;\,\varepsilon)$ exists such that $B(x;\,\varepsilon) \subseteq X$. Let X be a subset of $Y \subseteq R^n$; the *interior* of X relative to Y is then defined as being the set

$$\mathrm{Int}_Y\, X := \{x \in Y \mid \exists \varepsilon > 0 : B(x;\,\varepsilon) \cap Y \subseteq X\}.$$

A set X is said to be *open* in Y, if $\mathrm{Int}_Y X = X$.
A set $X \subseteq Y$ is said to be *closed* in Y if its complement, $Y \backslash X$, is open in Y.

If a set X is open or closed relative to R^n, the concepts are simply called open or closed, and the suffix R^n in the symbols omitted.

The intersection of all closed sets containing a set X is said to be the *closure* of X:

$$\begin{aligned} \mathrm{Cl}\, X &:= \bigcap \{B(X;\,\varepsilon) \mid \varepsilon > 0\} \\ &= \{x \mid \forall \varepsilon > 0 : B(x;\,\varepsilon) \cap X \neq \phi\}. \end{aligned}$$

It follows that a set X is closed if $\mathrm{Cl}\, X = X$. It can be proven that any intersection of a class of closed sets is closed, and that the union of a finite number of closed sets is closed.

The concepts of open and closed sets can be introduced equivalently in terms of convergence of vectors with respect to the Euclidean metric. Let $\{x_t\}$ be a sequence of vectors in R^n, $t = 1, 2, \ldots$ This sequence *converges* to a *limit point* x, if for any $\varepsilon > 0$, an integer t_ε exists such that $t > t_\varepsilon$ implies $d(x_t, x) < \varepsilon$. This is written: $x_t \to x$, or: $\lim_{t \to \infty} x_t = x$. A set is closed when it contains all its limit points.
The *boundary* of a set X is defined as being the set

$$\mathrm{Bnd}\, X := \mathrm{Cl}\, X \backslash \mathrm{Int}\, X.$$

An element of the boundary of X is called a *boundary point* of X. A set X

is *bounded* if *some* $\varepsilon > 0$ exist such that $X \subseteq B(x, \varepsilon)$, for some $x \in X$. A set X is said to be *compact* if it is closed and bounded.

Let $f: X \to Y$ be a multifunction; f is said to be *closed*, or *set-closed*, if $f(A)$ is closed in $R(f)$ whenever A is a closed set in X.

f is said to be *point-closed* if $f(x)$ is closed in $R(f)$ for every $x \in X$.

f is said to be *graph-closed* if the graph of f, $G(f)$ is a closed set in $X \times Y$.

f is said to be *open* if the set $f(A)$ is open whenever A is an open set in X.

Examples in two dimensions are given in figures 8.2.1–8.2.3.

Fig. 8.2.1. f is u.h.c.

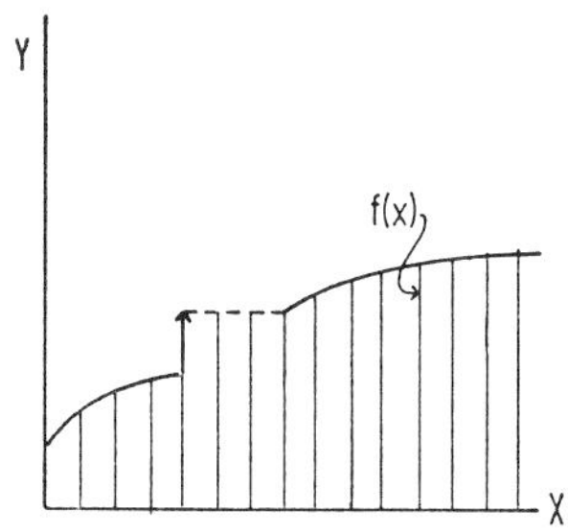

Fig. 8.2.2. f^{-1} is set-closed.

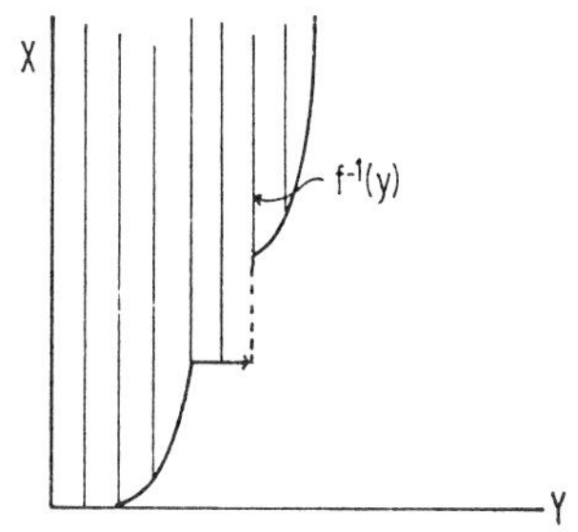

The multifunction f in fig. 8.2.1 is not point-closed, and is therefore neither closed nor graph-closed. The inverse f^{-1}, however, in fig. 8.2.2 is closed and is therefore also point-closed, but not graph-closed.

Let $f: X \to Y$ be a multifunction. f is said to be *upper hemi-continuous* (*u.h.c.*) if $f^{-1}(A)$ is a closed set in $R(f^{-1})$ whenever A is a closed set in Y.

f is said to be *lower hemi-continuous* (*l.h.c.*) if $f^{-1}(A)$ is an open set in $R(f^{-1})$ whenever A is open in Y.

f is said to be *continuous* if it is both u.h.c. and l.h.c.

Examples of an u.h.c. multifunction in two dimensions are given in figs. 8.2.1 and 8.2.3. It should be noted that neither inverse is u.h.c. (although the inverse of the multifunction in fig. 8.2.3 is graph-closed), because $f(A)$ is not closed in $R(f)$ for the unbounded closed set A (see also property 8.2.2.2 below).

Let $f: X \to Y$ be a function; then the continuity of f is equivalent to f being u.h.c. and equivalent to f being l.h.c.

Fig. 8.2.3. f is u.h.c.

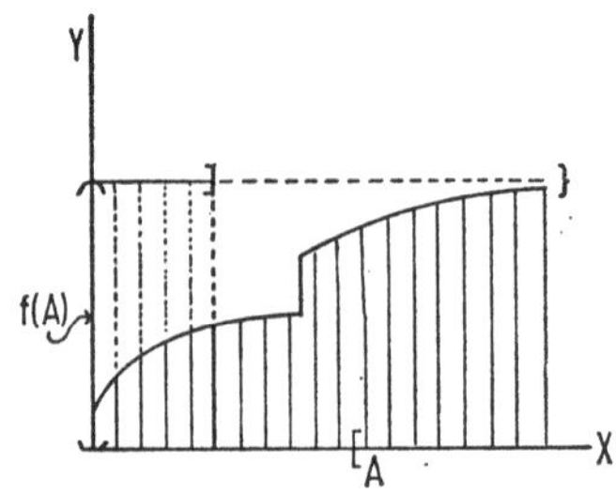

Fig. 8.2.4. f is l.h.c.

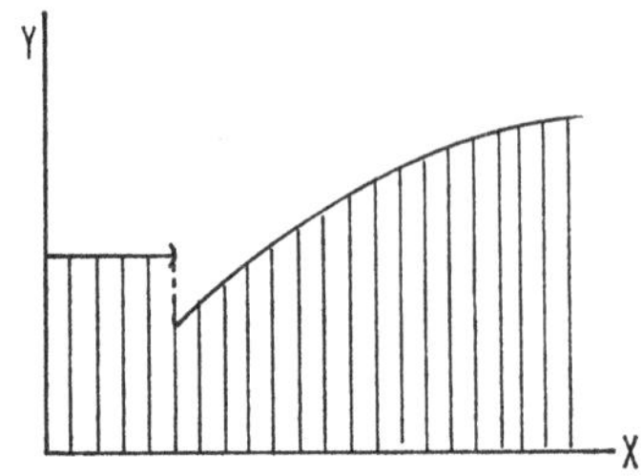

Property 8.2.1. (equivalent definitions)

1. A multifunction $f: X \to Y$ is *u.h.c.* if and only if whenever $x \in X$ and $U \subseteq Y$ is open with $f(x) \subseteq U$, an open set $V \subseteq D(f)$ with $x \in V$ exists, such that $f(V) \subseteq U$.
2. A multifunction $f: X \to Y$ is *l.h.c.* if and only if whenever $x \in X$, $y \in f(x)$ and U is an open set containing y, an open set V with $x \in V$ exists such that $f(z) \cap U \neq \phi$ for all $z \in V$.
3. A multifunction $f: X \to Y$ is *graph-closed* if and only if whenever $x \in X$, $y \in Y$ and $y \notin f(x)$, two open sets U and V with $y \in U$ and $x \in V$ exist such that $f(z) \cap U = \phi$ for all $z \in V$.
4. A multifunction $f: X \to Y$ is *l.h.c.* if and only if for all $x \in X$: $x_t \to x$ and $y \in f(x)$ imply that a sequence $\{y_t\}$ exists such that $y_t \in f(x_t)$ and $y_t \to y$.
5. A multifunction $f: X \to Y$ is *graph-closed* if and only if for all $x \in X$: $x_t \to x$, $y_t \to y$ and $y_t \in f(x_t)$ imply $y \in f(x)$.

The equivalence under 1. and 2. has been shown by Smithson (1972); 3. has been shown by Berge (1959, p. 111); 4. and 5. are equivalent formulations of 2. and 3.

Property 8.2.2. (graph-closed vs. u.h.c.)

Let $f: X \to Y$ be a multifunction.
1. If f is u.h.c. and point-closed, then f is graph-closed.
2. If Y is a compact set and f is graph-closed, then f is u.h.c.

This property has been shown by Nikaido (1968, p. 66).
If one defines upper hemi-continuity of a multifunction by requiring that its graph be closed (see Arrow and Hahn, 1971, p. 423), then this defini-

tion may contradict the definitions given above and also Berge's or Nikaido's definition of upper hemi-continuity at a point. This contradiction, however, is not possible if one defines upper hemi-continuity of a multifunction by requiring that the multifunction be graph-closed and have a compact range (Debreu, 1959, p. 17). This definition of Debreu is equivalent to Berge's definition, which requires that an u.h.c. multifunction be point – u.h.c. and point – compact (Berge, 1959; p. 109). This equivalence follows from property 8.2.2.

The following properties have been shown by Berge (1963, p. 113, 69 and 161):

Property 8.2.3. (*continuity preserving operations*)

The composition product, the union and the cartesian product of a family of l.h.c. (u.h.c., resp. point-compact) multifunctions is also l.h.c. (u.h.c., resp. point-compact). The sum of a family of l.h.c. (resp. u.h.c. and point-compact) multifunctions is again l.h.c. (resp. u.h.c. and point-compact).

Property 8.2.4. (*Maximum theorem*)

Let φ be a continuous real-valued function in Y and let $f: X \to Y$ be a point-compact and continuous multifunction such that $X = D(f)$; then the real-valued function $\mu(x) := \max \{\varphi(y) \mid y \in f(x)\}$ is continuous in X and the multifunction $g: X \to Y$, defined by $g(x) := \{y \mid y \in f(x), \varphi(y) = \mu(x)\}$, is point-compact and u.h.c.

Upper hemi-continuity of a multifunction is related to the well-known concept of upper semi-continuity of a function, but is not a generalization of this latter concept. In order to indicate the relation, the following closures of a real valued function can be defined: Let $f: X \to Y$ be a real valued function, i.e. $Y \subseteq R$. The *less-closure* of f is said to be the multifunction $f_- : X \to R$ defined by

$$f_-(x) := \{\lambda \in R \mid \lambda \leqq f(x)\}, \text{ for all } x.$$

The *more-closure* of f is said to be the multifunction $f_+ : X \to R$ defined by

$$f_+(x) := \{\lambda \in R \mid \lambda \geqq f(x)\}, \text{ for all } x.$$

Let $f: X \to Y$ be a real valued function; f is said to be *upper semi-continuous* (u.s.c.) on X if its less-closure $f_- : X \to R$ is a graph-closed multifunction. f is said to be *lower semi-continuous* (l.s.c.) on X if $-f$ is u.s.c. on X.

Property 8.2.5. (*equivalent definitions for u.s.c.*)

Let $f: X \to R$ be a real valued function. Then the following conditions are equivalent:

1. f is u.s.c. on X;
2. whenever $x \in X$ and $\varepsilon > 0$, an open set $V \subseteq X$ with $x \in V$ exists such that for all $v \in V, f(v) < f(x) + \varepsilon$;
3. f_- is u.h.c. on X;
4. f_+ is l.h.c. on X.

Fig. 8.2.5. An u.s.c. function with its less-closure

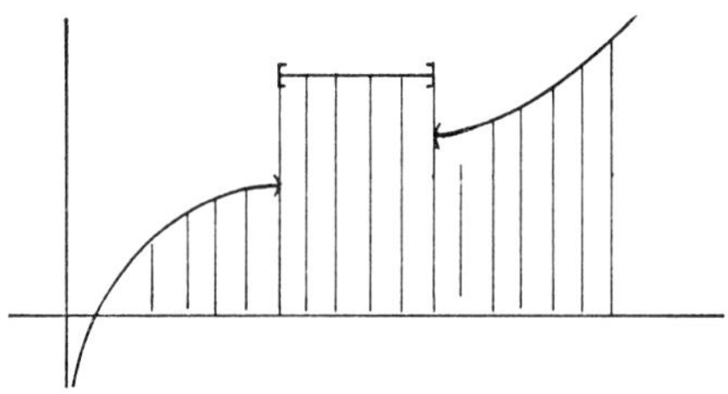

Proof: Equivalence of (1) and (2) follows from statements given by Berge (1959, p. 76) and Rockafellar (1970, p. 51).

Further (2) implies (3), since whenever $x \in X$ and $U := f_-(x) + \varepsilon \subseteq R$ which is an open set with $f_-(x) \subseteq U$, an open set $V \subseteq X$ with $x \in V$ exists such that $f_-(V) \subseteq U$. (3) follows from property 8.2.1.1. Statement (3) implies (1), through property 8.2.2.1, as f_- is point-closed, f being a function. Further, by definition $(-f)$ is l.s.c.; since $(-f)_- = -(f_+)$, it follows that (2) implies (4), for whenever $x \in X$ and $\varepsilon > 0$ an open set $V \subseteq X$ with $x \in V$ exists such that $-f(V) > -f(x) - \varepsilon$. This implies that $(-f)_-$ is l.h.c., because whenever $x \in X$, $\varepsilon > 0$, $y = -f(x)$ and $U := (-f)_-(x) + \varepsilon$, there exists an open set V with $x \in V$ such that whenever $z \in V$ there exists a $\delta > 0$ with $-f(z) - \delta = -f(x) - \varepsilon$, or $(-f)_-(z) \cap U \neq \phi$. Via property 8.2.1.2, $-(f_+)$ is u.h.c. and therefore also f_+. Finally, (4) implies (1) since for each open set in R, $(f_+)^{-1}$ is an open set in X, which is a sufficient condition for $-f$ being l.s.c. (Berge, 1959, p. 76) and f being u.s.c. □

An analogous property is valid for l.s.c. functions. A real valued function is continuous if it is both u.s.c. and l.s.c., or if it is l.h.c.

The set R^n is *partially preordered* by the following relation in R^n: let x_l and y_l be the l-th component of the vectors $x := (x_1, x_2, \dots, x_n)$ and $y := (y_1, y_2, \dots, y_n)$ in R^n and let N be the index set of n-numbers. Then
$x \leqq y$ means that $x_l \leqq y_l$, for every $l \in N$;
$x \leq y$ means that $x \leqq y$ and $x \ngeqq y$;
$x < y$ means that $x_l < y_l$, for every $l \in N$.
The *non-negative orthant*, R^n_+, is defined by $\{x \in R^n \mid x \geqq 0\}$.

Let x and y be different points in R^n; then the set $\{x + \lambda(y-x) \mid \lambda \in R\}$ is called a *line* through x and y. A subset X of R^n is said to be an *affine set* if $x + \lambda(y-x) \in X$ whenever $x \in X$, $y \in X$ and $\lambda \in R$ (see fig. 8.3.1).

An affine set X is called a *subspace* of R^n if it is empty or contains the origin. A subspace is closed under the operations of addition and scalar multiplication, i.e. $\lambda x + \mu y \in X$ whenever $x \in X$, $y \in X$, $\lambda \in R$ and $\mu \in R$.

Let x and y be different points in R^n; then the set $\{x + \lambda(y-x) \mid 0 \leqq \lambda\}$ is called a *half-line* from x through y with *direction* $(y-x)$. A subset X of R^n is said to be a *cone with vertex* x if $x + \lambda(y-x) \in X$ whenever $y \in X$ and $\lambda \geqq 0$

A cone with vertex 0, i.e. the origin, is just called a *cone*. A cone is closed under the operation of non-negative scalar multiplication, i.e. $\lambda x \in X$ whenever $x \in X$ and $\lambda \geqq 0$. A set X in R^n is said to be *aureoled* if it is closed under the operation of multiplication with scalar $\lambda \geqq 1$, i.e. $\lambda x \in X$ whenever $x \in X$ and $\lambda \geqq 1$ (see fig. 8.3.2).

Let x and y be different points in R^n; then the set $\{x + \lambda(y-x) \mid 0 \leqq \lambda \leqq 1\}$ is called a closed *line segment* and denoted by $[x, y]$. The open line segment $\{x + \lambda(y-x) \mid 0 < \lambda < 1\}$ is denoted by (x, y). A subset X of R^n is said to be *convex* if $x + \lambda(y-x) \in X$ whenever $x \in X$, $y \in X$ and $0 \leqq \lambda \leqq 1$. It should be noticed that each affine set is a convex set, but that the converse is not true.

A subset X of R^n is said to be *starred with vertex* x if $x + \lambda(y-x) \in X$ whenever $y \in X$ and $0 \leqq \lambda \leqq 1$. A starred set with vertex x may thus be seen as a generalization of a cone with the same vertex. A set X is convex if and only if X is starred with vertex x for every $x \in X$. A subset X of R^n with vertex 0, i.e. the origin, is called *starred.* A set X is starred if and only if $\lambda x \in X$ whenever $x \in X$ and $0 \leqq \lambda \leqq 1$ (see fig. 8.3.2). A convex set containing 0 is always starred.

A cone which is convex is called a *convex cone*. A set X is a convex cone if and only if it is closed under the operations of addition and non-negative scalar multiplication, i.e. $\lambda x + \mu y \in X$ whenever $x \in X$, $y \in X$, $\lambda \in R_+$ and $\mu \in R_+$. A cone X is said to be ***pointed*** if $x \in X$ implies that $-x \notin X$, whenever $x \neq 0$

Fig. 8.3.1. The affine hull of X

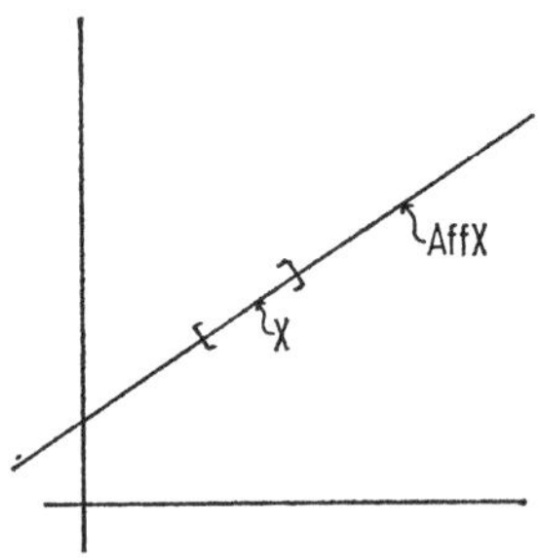

Fig. 8.3.2. The aureole closure $X \cup A$ and star closure $X \cup S$.

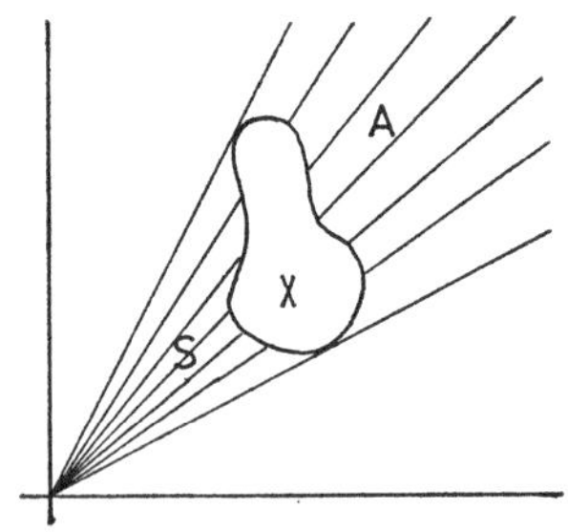

The following closure operations are defined on sets in R^n. Let X be a set in R^n. The intersection of the class of affine sets containing X is said to be the *affine hull* of X and is denoted by Aff X (see fig. 8.3.1);

$$\text{Aff } X = \{\sum_{i=1}^{n} \lambda_i x_i | x_i \in X, \lambda_i \in R, \sum_{i=1}^{n} \lambda_i = 1\}.$$

The intersection of the class of convex sets containing X is said to be the *convex hull* of X, denoted by Conv X;

$$\text{Conv } X = \{\sum_{i=1}^{n} \lambda_i x_i | x_i \in X, \lambda_i \in R_+, \sum_{i=1}^{n} \lambda_i = 1\}.$$

The *cone closure* of X is defined by

$$\text{Cone } X := \{\lambda x | x \in X, \lambda \geqq 0\}.$$

The *aureole closure* of X is defined by

$$\text{Aur } X := \{\lambda x | x \in X, \lambda \geqq 1\}.$$

The *star closure* of X is defined by

$$\text{Star } X := \{\lambda x | x \in X, 0 \leqq \lambda \leqq 1\}.$$

The *more closure* of X is defined by

$$\text{More } X := \{y | \exists x \in X : x \leqq y\}.$$

The *less closure* of X is defined by

$$\text{Less } X := \{y \mid \exists x \in X : y \leqq x\}.$$

The cone closure of X is the smallest cone containing X. A dual operation is the cone opening of X which defines the largest cone contained in X; this set is called the *interior cone* of X and denoted by

$$\text{Conint } X := \{x \mid \forall \lambda \geqq 0 : \lambda x \in X\}.$$

An important problem related with unbounded sets is the question how they behave in the infinite. It may be intuitively clear that, for some sets the interior cone, and for other sets the cone closure, give us information about the behavior of an unbounded set in the infinite. The following definition (given by Rockafellar, 1970, p. 61) adds more precision to this idea.

Let X be a non-empty convex set. Then X recedes in the direction of y, where $y \neq 0$, if and only if $x + \lambda y \in X$ for every $x \in X$ and $\lambda \geqq 0$. The direction of recession y indicates the direction in which the set X is unbounded (see fig. 8.3.3).

The *recession cone* of X, denoted by $0^+ X$, is said to be the set of all directions of recessions of X, including the origin:

$$0^+ X := \{y \in R^n \mid \forall \lambda \geqq 0, \forall x \in X : x + \lambda y \in X\}.$$

The notation indicates that the recession cone can be considered as the limit of a sequence of sets $\{\lambda X\}$ when $\lambda \to 0$ (see property 9.2.1). This notation will also prove fruitful in other definitions (see section 9.3). It is evident that X is bounded if and only if $0^+ X = \{0\}$.

Fig. 8.3.3. A convex set X and an aureoled convex set Y with their recession cones, $0^+ X$, *resp.* $0^+ Y$.

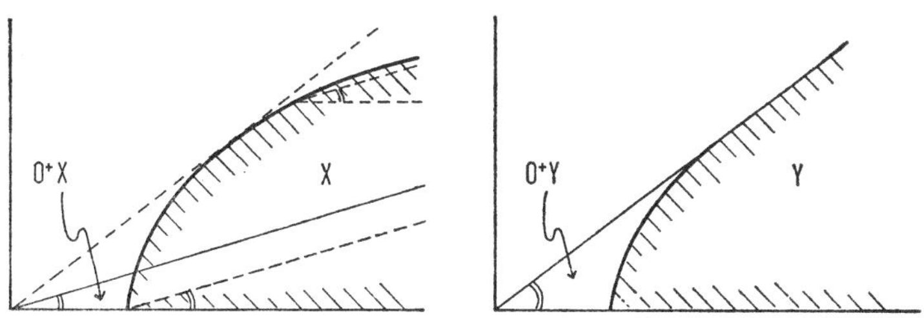

Another definition making the concept of behavior of a set at the infinite more precise, has been given by Debreu (1959, p. 22). Let X be any non-empty set; the *asymptotic cone* of X is defined (and denoted) by:

$$\text{Asc } X := \bigcap_{\lambda \geqq 0} \text{Cl Cone } \{x \in X \mid |x| \geqq \lambda\}.$$

If X is convex, then $0^+ X = \text{Asc } X$. The definition of a recession cone, rather than that of an asymptotic cone, will be used here.

Property 8.3.1. (*recession cones*)

Let X be a non-empty closed and convex set. Then:
1. $0^+ X = \text{Conint } X$, if X is starred;
2. $0^+ X = \text{Cl Cone } X$, if X is aureoled.

Proof
1. $0^+ X \subseteq \text{Conint } X$. Since X is starred, $0 \in X$. Then $y \in 0^+ X$ implies that $\forall \lambda \geqq 0 : 0 + \lambda y \in X$. Thus it follows that $y \in \text{Conint } X$.

$0^+ X \supseteq \text{Conint } X$. Suppose that $\text{Conint } X = \phi$; then X is bounded and $0^+ X = \{0\} \supseteq \phi$. Suppose that Conint X is not empty and choose $y \in \text{Conint } X$. Then $\lambda \geqq 0 \Rightarrow \lambda y \in X$. Let $x \in X$ and $\mu \in (0, 1)$; from the convexity of X it follows that $x + \mu(\lambda y - x) \in X$, for any $\lambda \geqq 0$. Define $\lambda := \kappa^2$ and $\mu := 1/\kappa$, with $\kappa > 1$. Then $x + \kappa y - x/\kappa \in X$. If $\kappa \to \infty$, the limit point of $x + \kappa y$ is an element of X and so is every convex combination of x and this limit point; or $x + \lambda y \in X$, for every $\lambda \geqq 0$. Therefore $y \in 0^+ X$.

2. $0^+ X \subseteq \text{Cl Cone } X$ (see fig. 8.3.3).
Choose $y \in 0^+ X$; if $y \in \text{Int } 0^+ X$, then a $\mu > 0$ exists such that $\mu y \in X$, or $y \in \text{Cone } X$. If for some $y \in 0^+ X$, no positive μ exists such that $\mu y \in X$, then $y \in \text{Cl Cone } X$. This follows from convexity of X, as for any $\bar{y} \in \text{Cone } X$, $x + \lambda \bar{y} \in X$ and $x + \lambda y \in X$ imply that $x + \lambda z \in X$, whenever $z \in [\bar{y}, y]$; therefore every open neighborhood of y has an element in Cone X.

$0^+ X \supseteq \text{Cl Cone } X$. Since X is aureoled, $\lambda x \in X$ whenever $x \in X$ and $\lambda \geqq 1$. Let $y \in \text{Cl Cone } X$; if $y \in X$, then $\lambda y \in X$, for $\lambda \geqq 1$, or $y + \lambda y \in X$, for $\lambda \geqq 0$. If $y \notin X$, then a $\mu > 0$ exists such that $x := y + \mu y \in X$. It follows that if $x \in X$ and $\lambda \geqq 0$, then $x + \lambda y \in X$. Therefore $y \in 0^+ X$. □

The following algebraic operations on the sets X and Y in a finite real Euclidean space are used.

Let a be a vector in R^n; then the *a-translation* of $X \subseteq R^n$ is said to be the set

$$X + \{a\} := \{x + a \mid x \in X\}.$$

Let λ be a scalar in R; then the *scalar multiplication* of X by λ is said to be the set.

$$\lambda X := \{\lambda x \mid x \in X\}.$$

Let $X \subseteq R^m$ and $Y \subseteq R^n$; then the *direct sum* of X and Y is said to be the set

$$X \oplus Y := \{z \in R^{m+n} \mid \exists x \in X, \exists y \in Y : z = (x, y)\}.$$

Let $X \subseteq R^n$ and $Y \subseteq R^n$; then *partial addition* of X and Y relative to the subspace R^m is said to be the set (see fig. 8.3.4)

$$X \# Y := \{(z_1, z_2) \in R^n \mid \exists (x_1, x_2) \in X, \exists (y_1, y_2) \in Y : \\ z_1 = x_1 = y_1 \in R^{n-m}, z_2 = x_2 + y_2 \in R^m\}.$$

The operation of partial addition of X and Y has as extremes $m = n$ and $m = 0$. Defined respectively:
addition: the sum of X and Y is said to be the set

$$X + Y := \{z \mid \exists x \in X, \exists y \in Y : z = x + y\}$$

intersection: the intersection of X and Y is said to be the set

$$X \cap Y := \{z \mid \exists x \in X, \exists y \in Y : z = x = y\}$$

Fig. 8.3.4. Addition and partial addition relative to R^m

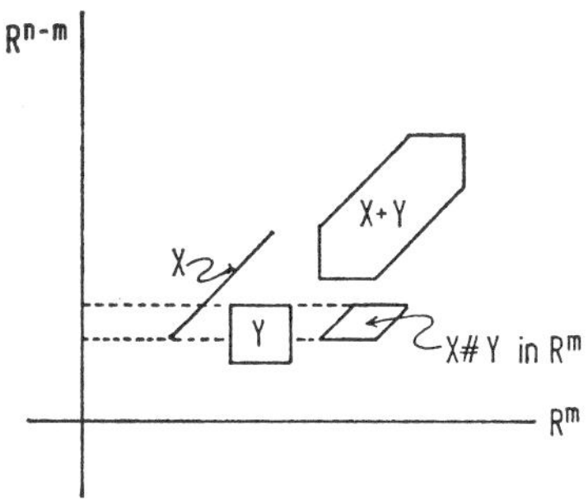

Finally, two algebraic operations are defined on convex sets X and Y. As they are narrowly related to addition and intersection of sets (see property 8.3.2), the operations are called convex addition and convex intersection.

Let X and Y be convex sets in R^n. Then:
convex addition of X and Y is defined (and denoted) by:

$$X \mathring{+} Y := \{z \mid \exists x \in X, \exists y \in Y, \exists \lambda \in [0, 1] : z = (1 - \lambda)x + \lambda y\}$$

convex intersection[1] of X and Y is defined (and denoted) by:

$$X \mathring{\cap} Y := \{z \mid \exists x \in X, \exists y \in Y, \exists \lambda \in [0, 1] : z = (1-\lambda)x = \lambda y\}$$

Fig. 8.3.5. (Convex) addition and intersection of X and Y

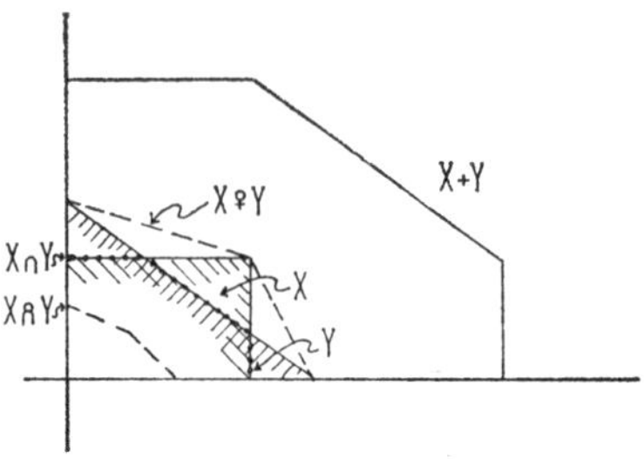

Property 8.3.2. (convex addition and intersection)

Let X and Y be two convex sets in R^n. Then

$$X \mathring{+} Y = \text{Conv}\,(X \cup Y).$$

Let X and Y be two convex cones containing the origin. Then

$$X \mathring{+} Y = X + Y,$$
$$X \mathring{\cap} Y = X \cap Y.$$

The proof of this property can be found in Rockafellar (1970, p. 80 and 22). The operation 'convex intersection' is called 'inverse addition' by Rockafellar and 'dual addition' by Weddepohl (1973). In order to indicate the analogy between operations on sets defined in this section and those operations on multifunctions to be defined in section 10.1., the term convex intersection is used here. This terminology also indicates the dual relation of addition and intersection in Eucledian vector spaces, as shown in the properties 9.3.4, 9.3.5, 10.4.6 and 10.4.7. It must be stressed, however, that just like addition, convex addition and convex intersection are both unions of an infinite number of convex sums, resp. intersections, i.e.

$$X \mathring{+} Y = \bigcup_{0 \leq \lambda \leq 1} \{(1-\lambda)X + \lambda Y\}$$

and

$$X \mathring{\cap} Y = \bigcup_{0 \leq \lambda \leq 1} \{(1-\lambda)X \cap \lambda Y\}.$$

[1] See also section 9.3.

Property 8.3.3. (convexity e.a. preserving operations)

The class of convex sets (resp. subspaces, cones) is closed under the operations of intersection, addition (direct, partial or convex), scalar multiplication and linear transformation. The class of convex sets (resp. affine sets) is also closed under the operation of translation.

This property has been shown by Berge (1959, p. 142) and Rockafellar (1970).

An affine set M is said to be *parallel* to an affine set L if a translation of M onto L exists. Each non-empty affine set is parallel to a unique subspace. The *dimension* of a non-empty affine set is defined as the dimension of the subspace parallel to it. The dimension of a convex set is defined as the dimension of the affine hull of that set.

Let X be a convex set in R^n. The *relative interior* of X, denoted by Rint X, is defined as being the interior of X relative to its affine hull.

Rint $X := \{x \in \text{Aff } X \mid \exists \varepsilon > 0 : B(x, \varepsilon) \cap \text{Aff } X \subseteq X\} = \text{Int}_{(\text{Aff } X)} X$. A convex set X is said to be *relatively open* if Rint $X = X$, e.g. the open line segment (a, b) in R^2. Any convex set X in R^n has the same dimension as Cl X or Rint X, as has been shown by Rockafellar (1970). He also proves that for any linear transformation $A : R^m \to R^n$, Rint $(AX) = A$ (Rint X) and Cl $(AX) \supseteq A$ (Cl X).

Let X and Y be convex sets in R^n. If both are (relatively) open, then their sum is open. It is not generally true, however, that the sum of two closed sets is closed (see fig. 8.3.6).

Fig. 8.3.6. An open sum of two closed sets

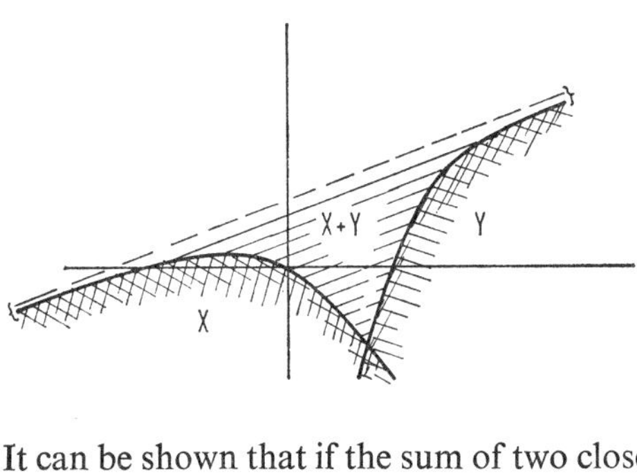

It can be shown that if the sum of two closed convex sets does not contain a line, then this sum is also closed. If a convex set X contains a subspace, then this subspace is also contained in the recession cone $0^+ X$. The largest subspace contained in $0^+ X$ is called the *linearity space* of X; its dimension is called the *lineality* of X. It is evident that if X does not contain a line, then the linearity space of X is equal to $\{0\}$ and the lineality of X is 0. The linearity space of X is equal to $(-0^+ X) \cap (0^+ X)$. The following property is shown by Rockafellar (1970):

Property 8.3.4. (*a closedness criterium*)

Let X be a non-empty closed convex set in R^n, and let A be a linear transformation from R^n into R^m. If for all $z \in 0^+ X$, $Az = 0$ implies that $z = 0$, then AX is closed.

Let X_i, $i \in M$, be a class of non-empty closed and convex sets in R^n such that $z_i \in 0^+ X_i$ and $\sum_i z_i = 0$ imply that z_i is an element of the linearity space of X_i for $i \in M$. Then $\sum_i X_i$ is closed, and $0^+ (\sum_i X_i) = \sum_i (0^+ X_i)$.

It should be noticed that the closedness condition in 8.3.4 is met when the sum of the recession cones $\sum_i (0^+ X_i)$ is a pointed cone. In that case, the recession cones are said to be positively semi-independent (Debreu, 1959).

As a final remark, it should be emphasized that the closedness condition in 8.3.4 is sufficient, but not necessary.

9. Sets and duality

9.1. SUPPORTING HYPERPLANES AND SEPARATING HYPERPLANES

Let p be a function of R^n into R; it is well known (see Fleming, 1965, p. 9) that p is linear if and only if real numbers $d_1, d_2, \ldots, d_n$ exist such that $p(x) = \sum d_i x_i = dx$, for every $x \in R^n$. The object d will be called a *covector* having components $d_1, d_2, \ldots, d_n$. Because a linear function p and a covector d have the same properties, they will be indicated here by the same symbol, such that $p(x) = px$, with p being identical to d.

Let $P(R^n, R)$ denote the set of all linear functions having domain R^n and values in R, and let the following operations be defined on these functions:

(addition)	$(p+q)(x) := p(x) + q(x)$,	for every $x \in R^n$;
(scalar multiplication)	$(\lambda p)(x) := \lambda p(x)$,	for every $x \in R^n$.

Then $P(R^n, R)$ is again a real Euclidean n-space (see 8.2), called the *dual space* of R^n and denoted by R^{n*}. The dual space R^{n*} has the same dimension as the primal space R^n. In fact they are isomorphic, and for many purposes, do not need to be distinguished, but this isomorphism is unnatural from several other points of view (Fleming, p. 287): more natural is the resemblance between R^{n**} and R^n.

The distinction between primal space and dual space is also important in economics, as the primal space is identified with the *quantity space* and the dual space with the *price space* or *valuation space*. Therefore, R^n and R^{n*} will always be distinguished, but not R^n and R^{n**}.

Finally, it must be stressed that all results between elements in R^n and R^{n*} are also valid for elements in R^{n*} and R^n.

Linear functions in R^n are thus elements (covectors) of the dual space, R^{n*}. They can, however, also be visualised or represented in the primal space by hyperplanes (see fig. 9.1.1).

A set in R^n is said to be a *hyperplane* if it is an affine set of dimension

Fig. 9.1.1. The inverse image of a function p with value α, and the covector p

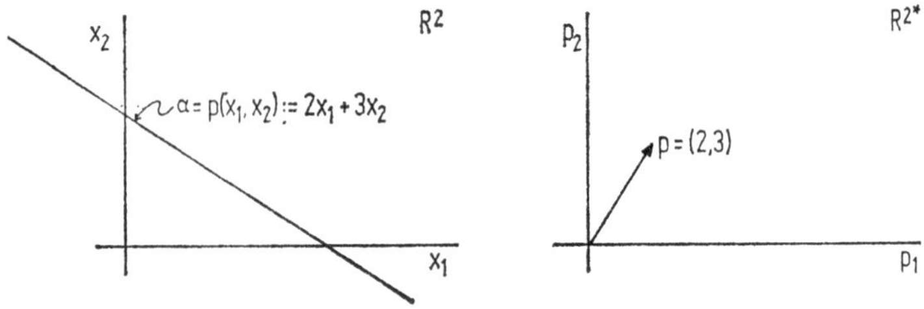

$(n-1)$. Let $p \in R^{n*}$ be a non-zero linear function and $\alpha \in R$ be a constant. Then the set

$$H(p;\alpha) := \{x \in R^n \mid px = \alpha\}$$

is a hyperplane in R^n. Conversely, every hyperplane may be represented in this way, with p and α being unique up to a common non-zero multiple (see Rockafellar, 1970). Whenever $\alpha \neq 0$, it is possible by appropriate scaling to choose a unique covector such that the value of the constant is equal to 1. Let $H(p;\alpha)$ be given and choose $\bar{p} := p/\alpha$; then $H(p;\alpha) = H(\bar{p}; 1)$. Therefore, a hyperplane not containing zero (such as the set in fig. 9.1.1), is uniquely represented by

$$H(p) := H(p; 1) = \{x \in R^n \mid px = 1\}.$$

With the hyperplane $H(p;\alpha)$ the sets

$$M_-(p;\alpha) := \{x \in R^n \mid px \leqq \alpha\} \quad \text{and}$$
$$M_+(p;\alpha) := \{x \in R^n \mid px \geqq \alpha\} = M_-(-p; -\alpha),$$

are associated. These are called the *lower closed half-space* and the *upper closed half-space* determined by $H(p;\alpha)$. Again, we define $M_+(p) := M_+(p; 1)$ and $M_-(p) := M_-(p; 1)$.

The unique correspondence between hyperplanes and linear functions (or covectors) is very fruitful when we take into account a special concept of hyperplanes: supporting hyperplanes. The concept of supporting hyperplanes may be considered as being a generalization of the concept of tangency used in classical analysis, and the corresponding covectors as a generalization of the differentials defined in classical analysis.

A hyperplane H in R^n is said to be a *supporting hyperplane* or a *support* for a set X if X is contained in some closed half-space M determined by H and, whenever V is an open set containing Cl X, V is not contained in M, i.e. $X \subseteq M$ and $V \nsubseteq M$ (see fig. 9.1.2). A hyperplane H is said to be a *properly-supporting hyperplane* for a set X if X is contained in some closed half-space M associated with H and H contains a point of X, i.e. $\subseteq XM$ and $X \cap H \neq \phi$ (see H_2 and H_3 in fig. 9.1.3).

It should be noted that what is here called properly-supporting, is often called supporting. The definition of a supporting hyperplane used here also includes the possibility of asymptotic support (see H_1, in fig. 9.1.2). A hyperplane H is called an *asymptotic support* for a set X if H does not contain a point of X but, whenever V is an open set containing Cl X, H contains a point of V, i.e. Cl $X \cap H = \phi$ and $V \cap H \neq \phi$. A supporting hyperplane for a closed set is thus either a properly supporting hyperplane or an asymptotic support.

Fig. 9.1.2. Supporting hyperplanes for X

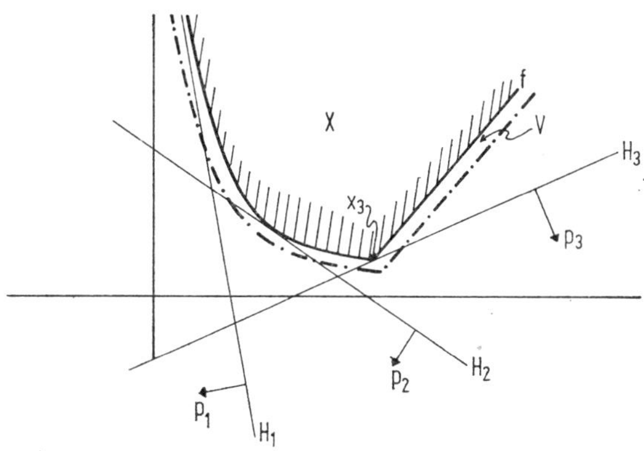

A related concept is the support function which assigns to some linear function p the supremum (or least upper bound) of the set of values given by p to elements of a convex set X.

The *support function* for a convex set X is the function $\sigma_x : R^{n*} \to R$, defined by $\sigma_x(p) := \sup \{px \mid x \in X\}$. For example: let X be the convex set in fig. 9.1.2; then $\sigma_x(p_3)$ is equal to the value of p_3 at the boundary point x_3 of X. It follows that a hyperplane $H(p; \alpha)$ is supporting for a set X if and only if $\sigma_x(p) = \alpha$. A hyperplane $H(p; \alpha)$ is properly-supporting for a set X if and only if $\sigma_x(p) = \alpha = \max \{px \mid x \in X\}$.

Another fruitful property of hyperplanes is based on the fact that a hyperplane separates R^n into two disjoint open convex sets: the two open half-spaces associated with the hyperplane.

Let X and Y be non-empty sets in R^n. A hyperplane H is said to *separate* X and Y if X is contained in one of the closed half-spaces associated with H and Y is contained in the other, i.e. $X \subseteq M_+$ and $Y \subseteq M_-$. The hyperplane H is called a *properly*-separating hyperplane if it separates X and Y and if neither X nor Y are contained in H. The hyperplane H is called a *strongly*-separating hyperplane if open sets $V \supseteq \mathrm{Cl}\, X$ and $W \supseteq \mathrm{Cl}\, Y$ exist such that $V \subseteq M_+$ and $W \subseteq M_-$. The hyperplane H is called a *supportingly*-separating hyperplane if it separates X and Y and if, for all open sets V and W with $\mathrm{Cl}\, X \subseteq V$ and $\mathrm{Cl}\, Y \subseteq W$, the intersection $V \cap W$ is not empty. If H is a supportingly-separating hyperplane for X and Y, then H is a supporting hyperplane for both X and Y (see fig. 9.1.3).

Fig. 9.1.3. A supportingly-separating hyperplane

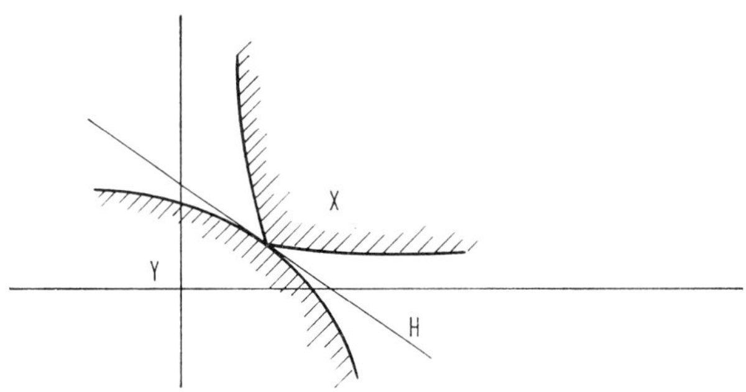

A hyperplane H strongly separating X and Y exists if and only if there exists a covector p such that $\sup\{px \mid x \in Y) < \inf\{px \mid x \in X\}$. There are many separation theorems, mostly based on the Hahn-Banach theorem. The following is essential in this context (see also Rockafellar, 1970, p. 97).

Property 9.1.1. (*separation theorem*)

Let X and Y be non-empty convex sets in R^n. In order that there be a hyperplane separating X and Y properly, it is necessary and sufficient that Rint X and Rint Y have no point in common.

The fact that a hyperplane separates R^n into two half-spaces is also used to define the following class of hyperplanes associated with a set X. A hyperplane is said to be a *bounding hyperplane* for X, if X is contained in some half-space associated with it. It is evident that every supporting hyperplane for X is a bounding hyperplane for X. The set of bounding hyperplanes for X generates the dual or polar set of X, as is shown in the next section.

9.2. POLAR SETS

Let X be a set in R^n. The set of bounding hyperplanes for X may be considered as being a dual set of X. For any non-zero constant real number α, there exists a unique correspondence between the bounding hyperplanes for X and a set of linear functions or covectors in R^{n*}. This set is called the *polar set of X relative to* α:

$$X_\alpha^* := \{p \in R^{n*} \mid X \subseteq M_-(p;\alpha) \quad \text{or} \quad X \subseteq M_+(p;\alpha)\}.$$

Since any intersection of closed half-spaces is a closed and convex set, it follows that $X_\alpha^* = (\text{Conv Cl } X)_\alpha^*$. An equivalent description is:

$$X_\alpha^* = \{p \in R^{n*} \mid L(p;\alpha) \cap \text{Rint}(\text{Conv } X) = \phi\}.$$

As the definition of the polar set X_α^* suggests, the set X_α^* can be partitioned into two sets, both of which are closed and convex.

The *upper polar set* relative to α is defined by

$$X_{\alpha+}^* := \{p \in R^{n*} \mid X \subseteq M_+(p;\alpha)\};$$

the *lower polar set* relative to α is defined by

$$X_{\alpha-}^* := \{p \in R^{n*} \mid X \subseteq M_-(p;\alpha)\}.$$

For any non-negative α, the upper polar set contains all covectors associated with hyperplanes separating X and $\{0\}$; the lower polar set contains all covectors associated with hyperplanes bounding for $X \cup \{0\}$ (see fig. 9.2.1).

If the value of α is not relevant for the problem considered, then α is assigned equal to 1 and not mentioned in the symbols and definitions:

$$\begin{aligned} X_+^* &:= \{p \in R^{n*} \mid X \subseteq M_+(p)\} \\ &= \{p \in R^{n*} \mid \text{for all } x \in X\colon px \geqq 1\}; \\ X_-^* &:= \{p \in R^{n*} \mid X \subseteq M_-(p)\} \\ &= \{p \in R^{n*} \mid \text{for all } x \in X\colon px \leqq 1\}; \\ X^* &:= X_+^* \cup X_-^*. \end{aligned}$$

Fig. 9.2.1. A set X and its polar set $X^* = X^*_+ \cup X^*_-$

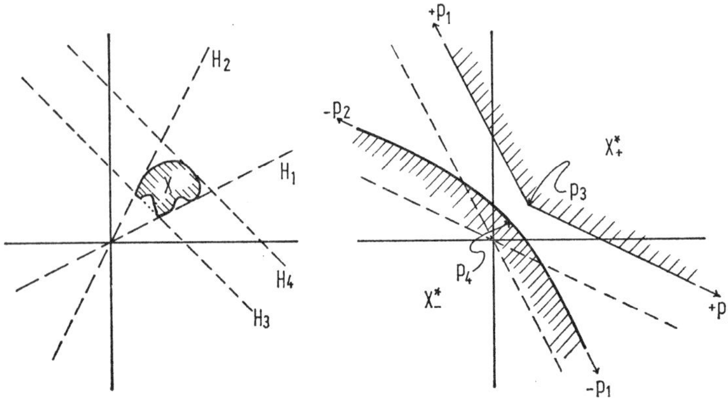

Another polarity operation is defined as follows: let X be a set in R^n. The set of covectors associated with all bounding hyperplanes for X which contain zero is said to be the *polar cone* of X:

$$
\begin{aligned}
X^0 &:= \{p \in R^{n*} \mid X \subseteq M_+(p;0) \quad \text{or} \quad X \subseteq M_-(p;0)\};\\
X^0_+ &:= \{p \in R^{n*} \mid \text{for all } x \in X\colon px \geqq 0\};\\
X^0_- &:= \{p \in R^{n*} \mid \text{for all } x \in X\colon px \leqq 0\}.
\end{aligned}
$$

The last two sets are respectively called, the *upper polar cone* and the *lower polar cone* (see fig. 9.2.2).

Fig. 9.2.2. A set X and its polar cones X^0_+ and X^0_-

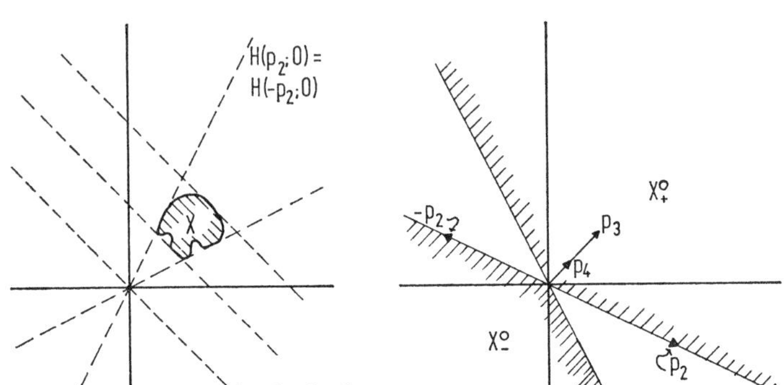

It is possible to express the polar cone as being a limit of a sequence of polar sets. The following definition of such a limit has been given (see Berge, 1959, p. 119):

Let $\{P_n\} := \{P_1, P_2, \ldots\}$ be a sequence of sets in R^{n*}, and let $N(p)$ be an open set containing p. The *lower limit* of the sequence $\{P_n\}$ is defined to be the set

$$\underline{\operatorname{Lim}}_{n\to\infty} P_n := \{p \mid \forall N(p), \exists n: \forall k \geqq n, P_k \cap N(p) \neq \phi\}.$$

The *upper limit* of that sequence is said to be the set

$$\overline{\operatorname{Lim}}_{n\to\infty} P_n := \{p \mid \forall N(p), \forall n, \exists k \geqq n: P_k \cap N(p) \neq \phi\}.$$

Clearly $\underline{\operatorname{Lim}}_{n\to\infty} P_n \subseteq \overline{\operatorname{Lim}}_{n\to\infty} P_n$. If the lower limit of a sequence equals the upper limit then this set is called the *limit* of a sequence of sets and denoted by $\operatorname{Lim}_{n\to\infty} P_n$. If the sequence is decreasing or increasing then the sequence admits a limit (which can be empty).

Property 9.2.1.

Let X be a set in R^n. Then

$$X^0_+ = \operatorname{Lim}_{\alpha\to 0} X^*_{\alpha+} \qquad \text{and} \qquad X^0_- = \operatorname{Lim}_{\alpha\to 0} X^*_{\alpha-}.$$

Proof. Let β be a scalar $0<\beta<1$ and k, an integer, be a power of β. Let $P_k := \beta^k X^*_+$. As $\alpha X^*_+ = X^*_{\alpha+}$, the sequence $\{P_n\} := (P_1, P_2, \ldots)$ is a subsequence of $\{X^*_{\alpha+}\}$.

Further, as $P_1 \subseteq P_2 \subseteq \ldots$, $\{P_n\}$ and $\{X^*_{\alpha+}\}$ are increasing sequences. As the lower limit of a sequence is contained in its upper limit, the property is proven if $X^0_+ \subseteq \overline{\operatorname{Lim}}\, P_n \subseteq \underline{\operatorname{Lim}}\, P_n \subseteq X^0_+$. Choose a $\bar{p} \in \overline{\operatorname{Lim}}\, P_n$ and assume that $\bar{p} \notin X^0_+$. Then an open neighborhood $N(\bar{p})$ exists such that $N(\bar{p}) \cap X^0_+ = \phi$; it follows that $\forall p \in N(\bar{p})$, $\exists \bar{x} \in X: p\bar{x} < 0$. Therefore, given any number n, it is true that for all $k \geqq n$, $P_k \cap N(\bar{p}) = \{p \mid \forall x \in X: px \geqq \beta^k\} \cap N(\bar{p}) = \phi$. This contradicts $\bar{p}$ being an element of $\overline{\operatorname{Lim}}\, P_n$.

Choose a $\bar{p} \in \overline{\operatorname{Lim}}\, P_n$; let the set $N(\bar{p})$ be such that $\forall n$, $\exists k \geqq n$: $\beta^k X^*_+ \cap N(p) \neq \phi$. As $l \geqq k \Rightarrow \beta^l X^*_+ \supseteq \beta^k X^*_+$, it follows that $\bar{p} \in \underline{\operatorname{Lim}}\, P_n$.

Choose a $\bar{p} \in X^0_+$ and assume that $\bar{p} \notin \underline{\operatorname{Lim}}\, P_n$. Then a neighborhood $N(\bar{p})$ exists such that for some $\tilde{p} \in N(\bar{p})$ it is true that $\tilde{p}x = \delta > 0$ when $x \in X$, and also such that $\forall n$, $\exists k \geqq n: P_k \cap N(\bar{p}) = \phi$. Choose $\bar{n}$ such that

$\beta^{\bar{n}} < \delta$. Then $P_{\bar{n}} \cap N(\bar{p}) = \phi$ contradicts the fact that $\tilde{p} \in \beta^{\bar{n}} X_+^* \cap N(\bar{p}) = P_{\bar{n}} \cap N(\bar{p})$, implied by $\bar{p} \in X_+^0$.

A analogous argument can be used for the lower polar sets. □

Property 9.2.2. (*upper polar sets*)

Let X and Y be any set in R^n. Then:

1. $p \in \text{Bnd } X_+^* \Leftrightarrow H(p)$ is a supporting hyperplane for X and separates the sets X and $\{0\}$;
   $p \in \text{Int } X_+^* \Leftrightarrow H(p)$ strongly separates X and $\{0\}$;
2. $X_+^* = \phi \Leftrightarrow 0 \in \text{Cl Conv } X$;
3. $X_+^* = (\text{Rint } X)_+^* = (\text{Cl } X)_+^* = (\text{Conv } X)_+^* = (\text{Aur } X)_+^*$;
4. X_+^* is a closed, convex and aureoled set, not containing 0;
5. $(X_+^*)_+^* = X$, when X is a closed, convex and aureoled set not containing the origin;
6. $X \subseteq Y \Rightarrow X_+^* \supseteq Y_+^*$;
7. $(X \cup Y)_+^* = X_+^* \cap Y_+^*$.

Proof (see Weddepohl, 1972):

1. By definition; see fig. 9.2.1 in which $H_3 := H(p_3)$ is a supporting hyperplane.

2. By definition: X and $\{0\}$ can not be separated by $\{x \mid px = 1\}$, for any p.

3. Let p be such that $\forall x \in X : px \geqq 1$. Then it is also true that $p\bar{x} \geqq 1$ if: $\bar{x} \in \text{Rint } X$; or $\bar{x} \in \text{Cl } X$; or $\bar{x} = \Sigma \lambda_i x_i$, for $x_i \in X$, $\lambda_i \geqq 0$ and $\Sigma \lambda_i = 1$; or, finally, if $\bar{x} = \lambda x$, for $x \in X$ and $\lambda \geqq 1$.

4. X_+^* is convex since $(1-\lambda)px + \lambda qx \geqq 1$, whenever $px \geqq 1$, $qx \geqq 1$, and $\lambda \in [0, 1]$, for all $x \in X$. Closedness follows from the contradiction which arises when $p \in \text{Cl } X_+^*$ and $p \notin X_+^*$; this implies that an $\bar{x} \in X$ exists such

Fig. 9.2.3.

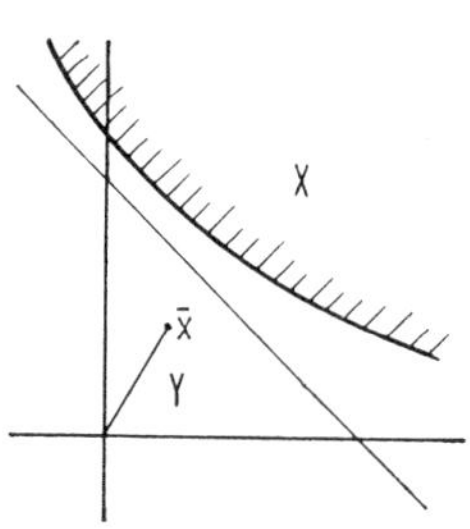

that $p\bar{x}<1$ and thus, for some sufficiently small $\varepsilon>0$, a $q\in B(p;\varepsilon)\cap X_+^*$ such that $q\bar{x}<1$ also. Finally, X_+^* is aureoled and does not contain zero, as $\lambda px \geqq 1$, when for all $x\in X: px\geqq 1$ and $\lambda\geqq 1$ (see also fig. 9.2.1).

5. $X\subseteq X_{++}^{**}$: choose $\bar{x}\in X$; then for all $p\in X_+^*: p\bar{x}\geqq 1$. It follows that $\bar{x}\in\{x\mid\forall p\in X_+^*: px\geqq 1\}=X_{++}^{**}$. $X\supseteq X_{++}^{**}$: assume that $\bar{x}\notin X$ and let $Y:=\{\lambda\bar{x}\mid\lambda\in[0,1]\}$. As X is closed, convex, aureoled and does not contain 0 and Y is closed, convex and does contain 0, there exists a hyperplane which strongly separates X and Y (property 9.1.1). Because $p\in X_+^*$ and $p\bar{x}<1$, it follows that $\bar{x}\notin X_{++}^{**}$ (see also fig. 9.2.3).

Fig. 9.2.4. $X\subseteq Y$ is equivalent to $X_+^*\supseteq Y_+^*$

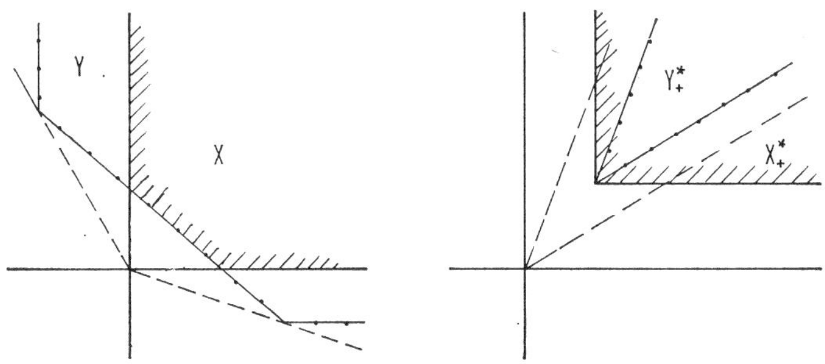

It may be noticed that in fig. 9.2.4 the set Y is not aureoled and therefore, although $Y\subseteq Y_{++}^{**}$, $Y\not\supseteq Y_{++}^{**}$.

6. If $p\in Y_+^*$ then $\forall y\in Y: py\geqq 1$. As $Y\supseteq X$, it is also true that $\forall y\in X$: $py\geqq 1$, i.e. $p\in X_+^*$. It follows that $Y_+^*\subseteq X_+^*$.

7. $(X\cup Y)_+^*\subseteq X_+^*\cap Y_+^*$ may be deduced from (6) as
follows: $X\subseteq(X\cup Y)\Rightarrow X_+^*\supseteq(X\cup Y)_+^*$;
also: $Y\subseteq(X\cup Y)\Rightarrow Y_+^*\supseteq(X\cup Y)_+^*$;
therefore: $(X_+^*\cap Y_+^*)\supseteq(X\cup Y)_+^*$.
To prove the converse, choose $p\in(X_+^*\cap Y_+^*)$; then $px\geqq 1$ when $x\in X$ and $px\geqq 1$ when $x\in Y$. Therefore $px\geqq 1$ when $x\in(X\cup Y)$, implying that $p\in(X\cup Y)_+^*$. □

Property 9.2.3. (*lower polar sets*)

Let X and Y be any set in R^n. Then:
1. $p\in\text{Bnd}\,X_-^*\Leftrightarrow H(p)$ is a supporting hyperplane for the set $X\cup\{0\}$;
   $p\in X_-^*\Leftrightarrow H(p)$ is a bounding hyperplane for $X\cup\{0\}$;

2. X^*_- is bounded $\Leftrightarrow$ $0 \in \text{Int Conv } X$;
3. $X^*_- = (\text{Rint } X)^*_- = (\text{Cl } X)^*_- = (\text{Conv X})^*_- = (\text{Star } X)^*_-$;
4. X^*_- is a closed, convex and starred set containing 0;
5. $(X^*_-)^*_- = X$, whenever X is a closed and convex set containing the origin;
6. $X \subseteq Y \Leftrightarrow X^*_- \supseteq Y^*_-$;
7. $(X \cup Y)^*_- = X^*_- \cap Y^*_-$.

The arguments of this proof are analogous to those given above for property 9.2.2. Compare also fig. 9.2.5.

Fig. 9.2.5. Lower polar sets of X and Y, with $X \subseteq Y$

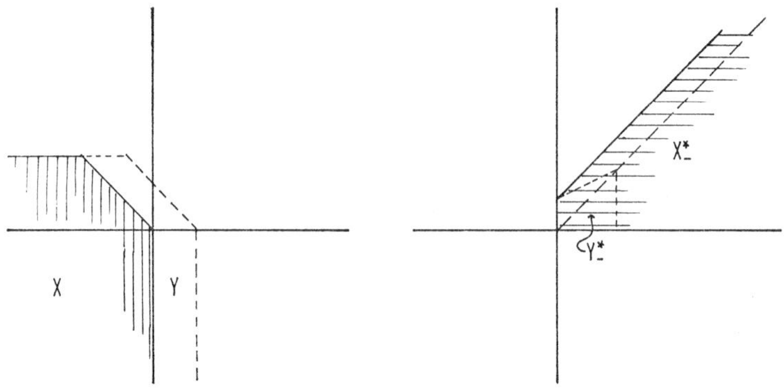

The difference between upper and lower cones is only a matter of signs, because $X^0_- = -X^0_+$. The properties of polar cones are well known and are given here for reference and comparison with similar properties of polar sets (see also fig. 9.2.2).

Property 9.2.4. (polar cones)

Let X and Y be any set in R^n. Then
1. $p \in \text{Bnd } X^0_-$ $\Leftrightarrow$ $H(p, 0)$ is a supporting subspace for X;
   $p \in X^0_-$ $\Leftrightarrow$ $H(p, 0)$ is a bounding subspace for X;
2. $X^0_- = \phi$ $\Leftrightarrow$ $0 \in \text{Int Conv } X$;
3. $X^0_- = (\text{Rint } X)^0_- = (\text{Cl X})^0_- = (\text{Conv } X)^0_- = (\text{Cone } X)^0_-$;
4. X^0_- is a closed convex cone;
5. $(X^0_-)^0_- = X$ whenever X is a non-empty closed convex cone;

6. $X \subseteq Y \Leftrightarrow X^0_- \supseteq Y^0_-$;
7. $(X \cup Y)^0_- = X^0_- \cap Y^0_-$.

In property 9.2.1 it was shown that a polar cone is a limit of a sequence of polar sets. It will be demonstrated here that this limit is also equal to the recession cone of the polar set, which recession cone is defined for convex sets (see 9.3) and indicates the directions in which the set recedes.

Property 9.2.5. (polar sets and polar cones)

Let X be any non-empty set in R^n. Then

1. $X^0_+ = 0^+(X^*_+)$;
2. $X^0_+ = \text{Cl Cone}(X^*_+) \supseteq X^*_+$, if $X^*_+ \neq \phi$;
3. $X^0_- = (\text{Cone } X)^*_- = 0^+(X^*_-)$;
4. $X^0_- = \text{Conint}(X^*_-) \subseteq X^*_-$;

Proof:
1. $X^0_+ \subseteq 0^+ X^*_+ = \{q \mid \forall \lambda \geqq 0, \forall p \in X^*_+ : p + \lambda q \in X^*_+\}$.
Choose any $q \in X^0_+$, $p \in X^*_+$ and $\lambda \geqq 0$; then for all $x \in X$: $\lambda q x \geqq 0$ and $px \geqq 1$. Addition gives $(p+\lambda q)x \geqq 1$, $\forall x \in X$, or $(p+\lambda q) \in X^*_+$. Thus $q \in 0^+ X^*_+$.
$X^0_+ \supseteq 0^+ X^*_+$: choose any $q \in 0^+ X^*_+$; then $\forall \lambda > 0$ and $\forall x \in X$: $(p+\lambda q)x \geqq 1$, or $(1/\lambda)px + qx \geqq (1/\lambda)$. Choose λ arbitrarily large, then $\forall x \in X: qx \geqq 0$. It follows that $q \in X^0_+$.

2. As X^*_+ is convex and aureoled, property 8.3.1.2 can be applied, from which statement 2 follows.

3. The equality $X^0_- = (\text{Cone } X)^*_-$ is shown as follows.
Choose some $p \in X^0_-$; then $\forall x \in X: px \leqq 0$. It follows that $\forall \lambda \geqq 0$: $p\lambda x \leqq 0 \leqq 1$. Therefore $\forall y \in \text{Cone } X: py \leqq 1$, implying that $p \in (\text{Cone } X)^*_-$.
Choose any $p \in (\text{Cone } X)^*_-$; then $\exists \lambda > 0$ such that $\forall x \in X: px \leqq (1/\lambda)$. Choose λ arbitrarily large, then $\forall x \in X: px \leqq 0$. It follows that $p \in X^0_-$.
The equality $X^0_- = 0^+(X^*_-)$ can be shown by arguments derived analogously from those given under (1).

4. As X^*_- is a closed, starred, convex and non-empty set, property 8.3.1.1 can be applied; from 3. the proposition follows. □

Polarity operations on sets carry over characteristics of a set X in the primal space to a set in the dual space. The information about a set transmitted from the primal space to the dual space is different for each polarity operation. For example: the boundedness of a set X is not carried over to the polar cone of X. If a polarity operation, applied twice to a set X, restores the original set, then we call this set X *reflexive* under the polarity operation. In this section properties of operations on some reflexive sets are given

Let X be a set in R^n; then
X is called *aureole-reflexive* if $(X^*_+)^*_+ = X$; i.e. if and only if X is a closed, convex and aureoled set, not containing 0;
X is said to be *star-reflexive* if $(X^*_-)^*_- = X$; i.e. if and only if X is a closed, convex and starred set;
X is said to be *cone-reflexive* if $(X^0_-)^0_- = (X^0_+)^0_+ = X$; i.e. if and only if X is a closed convex cone.

The equivalent conditions follow from properties 9.2.2.5, 9.2.3.5 and 9.2.4.5.

An important property of star-reflexive sets is the dual relation between the interior cone of the primal set and the cone closure of the polar set. (See also fig. 9.2.5).

Property 9.3.1. (*duality for cones of star-reflexive sets*)

Let X be a closed and convex set containing zero. Then
1. $(\text{Cone } X)^*_- = \text{Conint } (X^*_-) = 0^+(X^*_-)$;
2. $(\text{Conint } X)^*_- = \text{Cl Cone } (X^*_-) = (0^+ X)^0_-$.

Proof
1. The first equation is stated in properties 9.2.5.3 and 9.2.5.4.
2. Substitute X^*_- for X in the first line. Then

$$(\text{Cone } X^*_-)^*_- = \text{Conint } (X^{**}_{--}).$$

As both $(\text{Cl Cone } X^*_-)$ and X are star-reflexive, the lower polarity operation gives:

$$(\text{Cl Cone } X^*_-)^{**}_{--} = \text{Cl Cone } (X^*_-) = (\text{Conint } X)^*_-.$$

Considering properties 8.3.1.1 and 1. above, the proposition is proven. □

Property 9.3.2. (*duality for cones of aureole-reflexive sets*)

Let X be a closed, convex and aureoled set, not containing the origin. Then:

$$0^+(X_+^*) = \text{Cl Cone}\,(X_+^*) = (\text{Cl Cone}\,X)_+^0 = X_+^0 = (0^+X)_+^0\,.$$

Proof
From property 8.3.1.2, it follows that $(\text{Cl Cone}\,X)_+^0 = (0^+X)_+^0$;
from property 9.2.4.3, it follows that $(\text{Cl Cone}\,X)_+^0 = X_+^0$; and property 9.3.2 follow from property 9.2.5.1 and 9.2.5.2. □

Algebraic operations applied on sets generate new sets in R^n. The question arises whether or not a relation between the polar set of aggregated sets and the polar sets of the composing sets exists. It will be shown that this question can be answered in the affirmative.

In order to preserve closure of the convex sum, however, the definition given in 8.3 must be slightly refined. The following notation is introduced: Let X be any non-empty convex set. Then

$$\{\lambda X \mid \lambda \geqq 0^+\} := \{\lambda X \cup 0^+X \mid \lambda > 0\}.$$

The notation $\lambda \geqq 0^+$ thus means that λX is taken to be 0^+X, the recession cone, rather than $\{0\}$ if $\lambda = 0$.

The operation of *convex intersection* of two non-empty convex sets X_1 and X_2 is defined by:

$$X_1 \mathring{\cap} X_2 := \cup\{(1-\lambda)X_1 \cap \lambda X_2 \mid 0^+ \leqq \lambda \leqq 1\}.$$

This definition replaces the definition given in section 8.3 and preserves the closure of the sets. If X and Y are compact, both definitions coincide.

The operations of addition and intersection of reflexive sets generate more reflexive sets; this also holds for the operations of convex addition and convex intersection (see property 9.3.3). Examples are given in fig. 9.3.1 and 9.3.2. These figures also provide examples of property 9.3.4, in which, inter alia, it is stated that addition of sets in the primal space is equivalent (under weak conditions) to convex intersection of polar sets in the dual space.

The following interpretation may be given: the dual operation to addition of sets is addition of the bounding hyperplanes to those sets (see sections 9.1 and 9.2). The hyperplanes to be added are parallel sets, represented by covectors in the dual space, such as p_1 and p_2 in fig. 9.3.2. Therefore that covector p must be found, for which $H(p) = H(p_1) + H(p_2)$.

Because in the example given in fig. 9.3.2., $H(p_2) = 2H(p_1)$, it follows that $H(p) = 3H(p_1) = 3/2\ H(p_2) = H(1/3\ p_1) = H(2/3\ p_2)$, and that $p = 1/3\ p_1 = 2/3\ p_2$, p therefore belongs to the convex intersection of X^* and Y^*. Similar examples can be given for the other operations.

Fig. 9.3.1. Algebraic operations and polarity

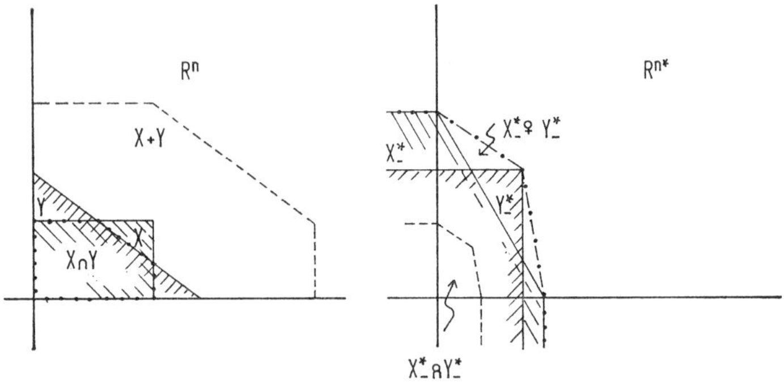

Property 9.3.3. (*operations on reflexive sets*)

Let X and Y be reflexive sets of the same type, i.e. either star-, aureole-, or cone-reflexive. Then

1. $X \cap Y$;
2. $X \mathring{\cap} Y$;

   are reflexive sets of the same type. If $X + Y$ is contained in a pointed cone, then

3. $X + Y$;
4. $X \mathring{+} Y$;

   are also reflexive sets of the same type.

Proof
Assume that X and Y are star-reflexive, i.e. that they are closed, convex and contain 0. According to section 8.2 and property 8.3.3, the intersection and convex-intersection of both sets are closed and convex. From property 8.3.3 it follows that the sum and convex sum are convex. The condition for addition and convex-addition also meets the closedness criterion in property 8.3.4. As the origin is evidently an element of all sets, property 9.3.3 follows. Analogous arguments can be given for the other types of reflexive sets. □

Property 9.3.4. (*polar operation and aggregated reflexive sets*)

Let X and Y be either star-reflexive or aureole-reflexive sets and let the operation * denote the lower-, resp. upper, polarity operation. Then:

1. $(X+Y)^* = X^* \mathring{\cap} Y^*$;
2. $(X\cap Y)^* = X^* \mathring{+} Y^*$, if $\operatorname{Int} 0^+(X\cap Y) \neq \phi$;
3. $(X \mathring{+} Y)^* = X^* \cap Y^*$;
4. $(X \mathring{\cap} Y)^* = X^* + Y^*$, if $\operatorname{Int} 0^+(X\cap Y) \neq \phi$;
5. $(\lambda X)^* \quad = \lambda^{-1} X^*$, for $\lambda > 0$.

Proof. All proofs will be given for aureole-reflexive sets. Analogous arguments can be given for star-reflexive sets (see fig. 9.3.1).

Fig. 9.3.2. $(X+Y)^* = X^* \mathring{\cap} Y^*$; *addition and convex intersection*

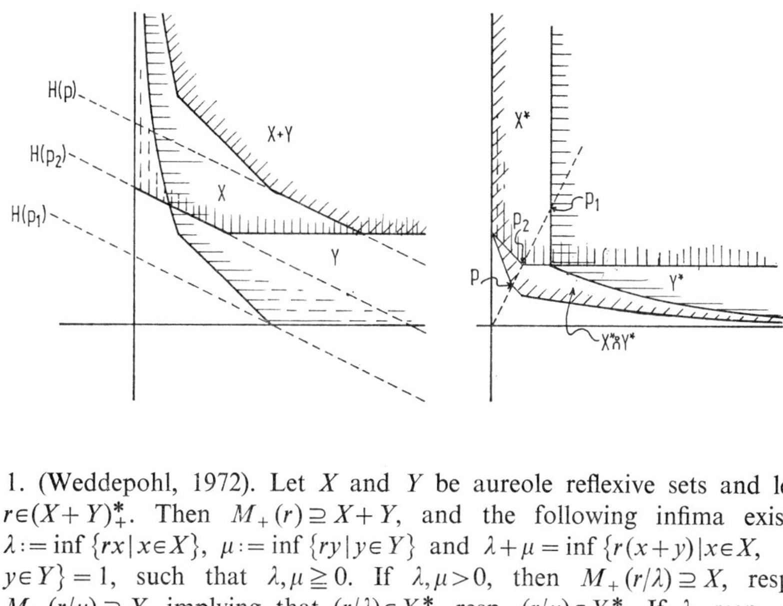

1. (Weddepohl, 1972). Let X and Y be aureole reflexive sets and let $r \in (X+Y)^*_+$. Then $M_+(r) \supseteq X+Y$, and the following infima exist: $\lambda := \inf\{rx \mid x \in X\}$, $\mu := \inf\{ry \mid y \in Y\}$ and $\lambda + \mu = \inf\{r(x+y) \mid x \in X, y \in Y\} = 1$, such that $\lambda, \mu \geqq 0$. If $\lambda, \mu > 0$, then $M_+(r/\lambda) \supseteq X$, resp. $M_+(r/\mu) \supseteq Y$, implying that $(r/\lambda) \in X^*_+$, resp. $(r/\mu) \in Y^*_+$. If λ, resp. μ, is zero, then $M_+(r;0) \supseteq X$, resp. $M_+(r;0) \supseteq Y$, implying that $r \in 0^+ X^*_+ = X^0_+$, resp. $r \in 0^+ Y^*_+ = Y^0_+$, according to property 9.2.5.1. Therefore $r \in (\lambda X^*_+ \cap \mu Y^*_+)$, for $\lambda, \mu \geqq 0^+$ and $\lambda + \mu = 1$.

Conversely, let $r = \lambda p = \mu q$, for $p \in X^*_+$, $q \in Y^*_+$ and $\lambda, \mu \geqq 0^+$, with $\lambda + \mu = 1$. Then for all $x \in X: px \geqq 1$, or $rx \geqq \lambda$, and for all $y \in Y: qy \geqq 1$, or $ry \geqq \mu$. Addition gives $r(x+y) \geqq 1$, for all $x \in X$ and $y \in Y$, implying that $r \in (X+Y)^*_+$.

4. As X^*_+ and Y^*_+ are aureole-reflexive, the sum of X^*_+ and Y^*_+ is aureole-reflexive if it is contained in a pointed cone (see property 9.3.3.3). In that case we can apply (1) above, from which follows:

$(X^*_+ + Y^*_+)^*_+ = (X^*_+)^*_+ \overset{\circ}{\cap} (Y^*_+)^*_+ = X \overset{\circ}{\cap} Y$, implying the statement after application of the upper polar operation. The sum $(X^*_+ + Y^*_+)$ is contained in Cone (X^*_+) + Cone(Y^*_+), which is shown to be a pointed cone as follows: from properties 9.2.5.2, 8.3.2 and 9.2.4 it can be seen that Cone(X^*_+)+Cone$(Y^*_+) \subseteq (X)^0_+ + (Y)^0_+ =$

$= \text{Conv}(X^0_+ \cup Y^0_+) \subseteq (X^{0\,0}_{+\,+} \cap Y^{0\,0}_{+\,+})^0_+ = (X \cap Y)^0_+$;

$[0^+(X \cap Y)]^0_+ = (X \cap Y)^0_+$ follows from property 9.3.2. Assume that $\exists r \in (\text{Cl Cone } X^*_+ + \text{Cl Cone } Y^*_+)$ such that $r \neq 0$ and $-r \in (\text{Cl Cone } X^*_+ + \text{Cl Cone } Y^*_+)$; then $\forall z \in 0^+(X \cap Y)$, $rz \geqq 0$ and $rz \leqq 0$. This contradicts the fact that Int $0^+(X \cap Y) \neq \phi$ and $\exists \delta > 0$ such that $(z + \delta r) \in 0^+(X \cap Y)$, implying that $r(z + \delta r) \neq 0$.

2. As X^*_+ and Y^*_+ are aureole-reflexive and as $(X^*_+ + Y^*_+)$ is contained in a pointed cone (see the argument under (4)), in which case $(X^*_+ \overset{\circ}{+} Y^*_+)$ is also aureole-reflexive, the application of the polarity operation on

$$(X^*_+ \overset{\circ}{+} Y^*_+)^*_+ = X^{**}_{++} \cap Y^{**}_{++} = X \cap Y,$$

results in statement 2 under property 9.3.4.

3. Since $(X \overset{\circ}{+} Y) = \text{Conv}\ (X \cup Y)$, and according to 9.2.2.3,

$$[\text{Conv}(X \cup Y)]^*_+ = (X \cup Y)^*_+,$$

property 9.3.4.3 follows from property 9.2.2.7.

5. Choose $p \in (\lambda X)^*_+$; then for all $y \in \lambda X$: $py \geqq 1$. Define $x := y/\lambda$; then for all $x \in X$: $\lambda px \geqq 1$, or $\lambda p \in X^*_+$ and $p \in \lambda^{-1} X^*_+$. □

The last property of this section follows directly from properties 8.3.2 and 9.3.4:

Property 9.3.5. (*polar operation and aggregated cones*)

Let X and Y be closed convex cones. Then:

1. $(X + Y)^0_- = X^0_- \cap Y^0_-$;
2. $(X \cap Y)^0_- = X^0_- + Y^0_-$, if Int $0^+(X \cap Y) \neq \phi$.

Let X be an aureole-reflexive set and Y be a star-reflexive set (this denotation will be used throughout this section); suppose that X and Y are separated by a hyperplane. The question arises: what behavior do the polar sets X_+^* and Y_-^* have relative to separation?

Apart from boundary cases, the answer is that no hyperplane separating X_+^* and Y_-^* exists if and only if there does exist a hyperplane strongly separating X and Y; and that a hyperplane supportingly separating X_+^* and Y_-^* exists if and only if there exists a hyperplane supportingly separating X and Y (see property 9.4.3).

This result is restricted by the behavior of certain unbounded sets under a polarity operation. It will therefore be required that the sets X and Y have no direction of recession in common, i.e. that there does not exist a $d \neq 0$ such that $d \in 0^+ X \cap 0^+ Y$. A pair of sets (X, Y) is said to have a *common orientation* if they have a direction of recession in common, i.e. if $0^+ X \cap 0^+ Y \supset \{0\}$. See fig. 9.4.1. The counterpart of this property in the dual space is given by the following theorem:

Property 9.4.1. (*common orientation and polarity*)

Let X be aureole-reflexive and Y be star-reflexive. Then X and Y have a common orientation if and only if X_+^* and Y_-^* are separated by a subspace (of dimension $n-1$).

Proof

Let (X, Y) be a pair of sets with common orientation and let some non-zero d be an element of $(0^+ X \cap 0^+ Y)$. Then for all $x \in X$ and $\lambda \geqq 0 : x + \lambda d \in X$. Choose any $q \in X_+^*$; then for all $x \in X : qx \geqq 1$. It is therefore also true that for all $x \in X$, $\lambda \geqq 0 : q(x+\lambda d) \geqq 1$. It will be shown that for all $\lambda \geqq 0 : q\lambda d \geqq 0$, or $q \in M_+(d; 0)$. Suppose that $qd = \beta < 0$ and for some $x \in X : qx = \alpha \geqq 1$. Then a $\lambda \geqq 0$ exists such that $qx + q\lambda d = \alpha + \lambda\beta < 1$, which contradicts the statement above. Therefore $X_+^* \subseteq M_+(d; 0)$. From a similar argument it follows that $Y_-^* \subseteq M_-(d; 0)$. Thus X_+^* and Y_-^* are separated by a hyperplane containing zero, $H(d; 0)$.

Conversely, let $H(d; 0)$ separate X_+^* and Y_-^* such that $X_+^* \subseteq M_+(d; 0)$ and $Y_-^* \subseteq M_-(d; 0)$. Let $L := \{\lambda d \mid \lambda \in R\}$ be a line with direction d and assume that L and X have no common orientation, i.e. $L \cap 0^+ X = \{0\}$.

From property 9.3.2 it follows that Cl Cone $X_+^* = (0^+ X)_+^0$. It is also true that $L_+^0 = M_+(d; 0)$ and $[\{0\}]_+^0 = R^{n*}$. As $L \cap 0^+ X = \{0\}$, by

Fig. 9.4.1. A pair of sets (X, Y) with common orientation

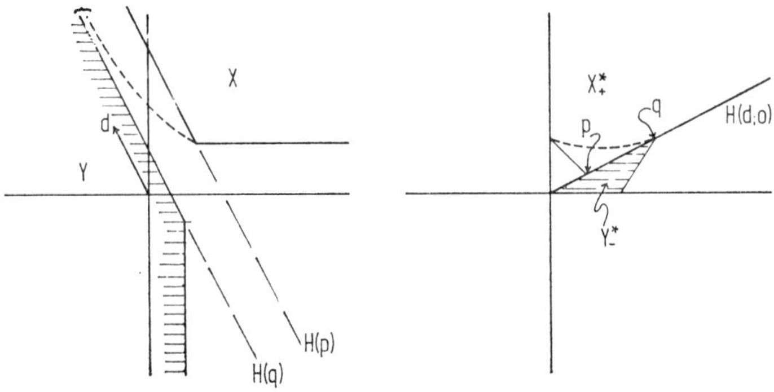

assumption, it follows from property 9.3.5 that $[L \cap 0^+ X]_+^0 = L_+^0 + (0^+ X)_+^0 = M_+(d; 0) + \text{Cl Cone } X_+^* = R^{n*}$. This clearly contradicts the fact that $X_+^* \subseteq M_+(d; 0) \subset R^{n*}$. A similar argument can be applied to show that L and Y have a common orientation. Therefore X and Y have a common orientation. □

The following theorem gives the relation between separation of X and Y in the primal and dual space without any restriction on X or Y.

Property 9.4.2. (separation and duality)

Let X be an aureole-reflexive set and Y be a star-reflexive set. Then:

1. a hyperplane strongly separating X and Y exists if and only if: either (a) there does not exist a hyperplane separating X_+^* and Y_-^* or (b) if a hyperplane containing zero exists supportingly separating X_+^* and Y_-^* then $\text{Rint}(X_+^* \cap Y_-^*) \neq \phi$.
2. (a) a hyperplane supportingly separating X and Y exists and (b) Rint $(X \cap Y) = \phi$ whenever the supportingly separating hyperplane contains zero, if and only if: (a) a hyperplane supportingly separating X_+^* and Y_-^* exists and (b) Rint $(X_+^* \cap Y_-^*) = \phi$ whenever the supportingly separating hyperplane contains zero.

Proof

1. Let X and Y be strongly separated; then there exists a p such that for some open sets V and W is true: $M_+(p) \supseteq V \supset X$ and $M_-(p) \supseteq W \supset Y$. Choose $\mu \neq 1$ such that for $q := \mu p$, $M_+(q) \supseteq X$ and $M_-(q) \supseteq Y$. This μ exists since both V and W are open sets (see fig. 9.4.2). As $M_+(p) \supseteq$

$M_+(q) \supseteq X$ and $M_-(q) \supseteq M_-(p) \supseteq Y$, the line segment $[p, q]$ is a subset of both $X_+^* \cap Y_-^*$ and the subspace $H := \{r \in R^{n*} \mid \exists \lambda \in R : r = \lambda p\}$.

Suppose that X and Y have a common orientation; according to property 9.4.1, X_+^* and Y_-^* are separated by a subspace, which contains the subspace H. Because Rint $(X_+^* \cap Y_-^*) \supseteq (p, q) \neq \phi$, condition (b) is met (see fig. 9.4.1).

Suppose that X and Y have no common orientation (see fig. 9.4.2). Then, according to property 9.4.1, no separating subspace for X_+^* and Y_-^* exists.

Fig. 9.4.2. A pair of strongly separated sets and its polar sets

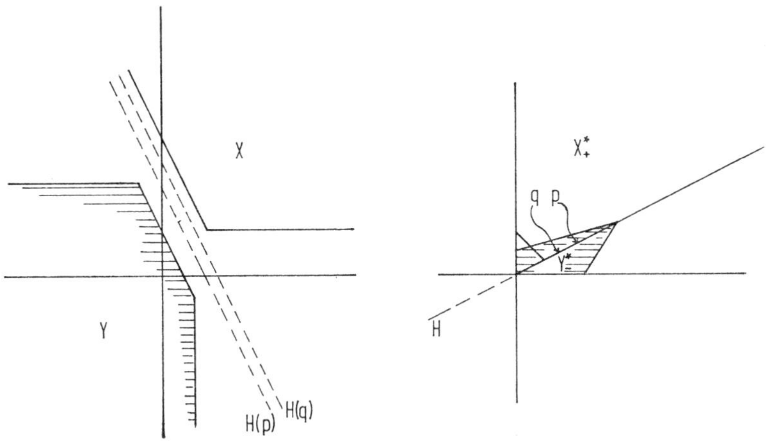

But there is certainly also no separating hyperplane $H(z)$ not containing zero. For, assuming that $M_+(z) \supseteq X_+^*$ and $M_-(z) \supseteq Y_-^*$, then according to property 9.2.3.6 $\{\lambda z \mid \lambda \geqq 1\} = [M_+(z)]_+^* \subseteq X$ and $\{\lambda z \mid 0 \leqq \lambda \leqq 1\} = [M_-(z)]_-^* \subseteq Y$, implying that $z \in (X \cap Y)$. This contradicts X and Y being strongly separated.

Conversely, let (a) there not exist a hyperplane separating X_+^* and Y_-^*. Then the separation property 9.1.1 states that there exists a $p \in (\text{Rint } X_+^* \cap \text{Rint } Y_-^*)$; let p be such that $M_+(p) \supseteq X$. As $p \in \text{Rint } X_+^*$, a $q \in X_+^*$ exists such that $M_+(p) \supset M_+(q) \supseteq X$, and an open set $V \supseteq \text{Cl } X$ exists such that $M_+(p) \supseteq V$. It also follows that an open set $W \supseteq \text{Cl } Y$ exists such that $M_-(p) \supseteq W$. Therefore $H(p)$ strongly separates X and Y.

Finally, let (b) a subspace separating X_+^* and Y_-^* exist such that Rint $(X_+^* \cap Y_-^*) \neq \phi$ (see fig. 9.4.1). Let $r \in \text{Rint}(X_+^* \cap Y_-^*)$ and $M_+(r) \supseteq X$. Then $\lambda < 1$ and $\mu > 1$ exist such that $p = \lambda r$ and $q = \mu r$, and $(p, q) \subseteq$

Rint $(X_+^* \cap Y_-^*)$. Since $M_+(r) \supset M_+(p) \supseteq X$ and $M_-(r) \supset M_-(q) \supseteq Y$, it follows that $H(r)$ strongly separates X and Y.

2. Define the relevant propositions as follows:
$a :=$ a hyperplane strongly separating X and Y exists;
$b :=$ a hyperplane separating X and Y exists;
$c :=$ if a subspace supportingly separating X and Y exists then Rint $(X \cap Y) = \phi$;
$d :=$ a hyperplane supportingly separating X and Y exists.
The negation is indicated by a bar, e.g. $\bar{a}$, and the proposition for X_+^* and Y_-^* by an asterisk, e.g. $a*$.

The statement to be proven is: d and $c \Leftrightarrow d*$ and $c*$.
The statement (1) says: $a \Leftrightarrow \bar{b}*$ or $\bar{c}*$;
from which follows: $\bar{a} \Leftrightarrow b*$ and $c*$,
and b and $c \Leftrightarrow \bar{a}*$.
Therefore: $\bar{a}$ and b and $c \Leftrightarrow \bar{a}*$ and $b*$ and $c*$.
Because it is true by definition that: $d \Leftrightarrow \bar{a}$ and b, the statement follows. □

Fig. 9.4.3. Supportingly separating hyperplanes

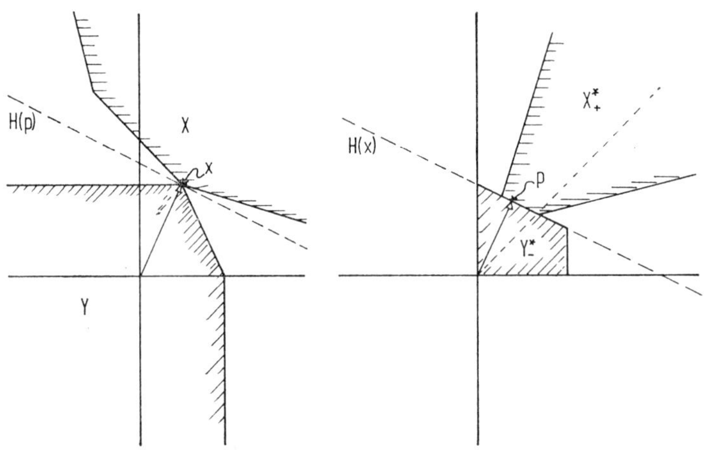

If the sets X and Y are slightly restricted, requiring that they do not have a common orientation or a common direction of recession, then the separation and duality property is reduced to the following (see also figs. 9.4.2 and 9.4.3):

Property 9.4.3. (*separation and duality*)

Let X be an aureole-reflexive set and Y be a star-reflexive set such that $0^+X \cap 0^+Y = \{0\}$ and there does not exist a hyperplane containing zero supportingly separating X and Y. Then:

1. A hyperplane strongly separating X and Y exists if, and only if, a hyperplane separating X^*_+ and Y^*_- does not exist;
2. A hyperplane supportingly separating X and Y exists if, and only if, a hyperplane supportingly separating X^*_+ and Y^*_- exists.

Proof

This property follows directly from property 9.4.2, since the conditions (b) under (1) and (2) in that property are ruled out by the restriction that there be no subspace separating either X and Y, or X^*_+ and Y^*_- (see property 9.4.1). □

The separation properties of sets can equivalently be expressed in intersection properties of sets, as is done in what follows:

Property 9.4.4. (*intersection and duality*)

Let X be an aureole-reflexive set and Y be a star-reflexive set such that $0^+X \cap 0^+Y = \{0\}$ and there be no $(n-1)$ dimensional subspace supportingly separating X and Y. Then:

1. $X \cap Y = \phi \quad \Leftrightarrow \text{Rint } X^*_+ \cap \text{Rint } Y^*_- \neq \phi;$
2. $\left.\begin{array}{l} X \cap Y \neq \phi \text{ and} \\ \text{Rint } X \cap \text{Rint } Y = \phi \end{array}\right\} \Leftrightarrow \left\{\begin{array}{l} X^*_+ \cap Y^*_- \neq \phi \text{ and} \\ \text{Rint } X^*_+ \cap \text{Rint } Y^*_- = \phi. \end{array}\right.$

If both X and Y have dimension n and do not contain a line, then:

3. $X \cap Y = \phi \quad \Leftrightarrow \text{Int}\,(X^*_+ \cap Y^*_-) \neq \phi;$
4. $\left.\begin{array}{l} X \cap Y \neq \phi \text{ and} \\ \text{Int}\,(X \cap Y) = \phi \end{array}\right\} \Leftrightarrow \left\{\begin{array}{l} X^*_+ \cap Y^*_- \neq \phi \text{ and} \\ \text{Int}\,(X^*_+ \cap Y^*_-) = \phi. \end{array}\right.$

Proof

According to the separation property 9.1.1, the non-existence of a hyperplane separating X and Y is equivalent to Rint $X \cap$ Rint $Y \neq \phi$.

It can also be shown that a hyperplane strongly separating X and Y exists if, and only if, $X \cap Y = \phi$ (see also Weddepohl, 1973):

Assume that an open set U exists such that for some hyperplane $H(p): M_+(p) \supseteq U \supset X$ and $M_-(p) \supseteq Y$. Then $X \cap Y = \phi$.

Conversely, let $X \cap Y$ be empty; then there exists a hyperplane separating X and Y, and some closed set $V := B(Y; \varepsilon)$ such that $V \cap X \neq \phi$ (see fig. 9.4.4). As $0^+ Y = 0^+ V$ and $0^+ X \cap 0^+ Y = \{0\}$, it follows that $0^+ X \cap 0^+ V = 0^+ (X \cap V) = \{0\}$. Therefore $X \cap V$ is compact and a hyperplane $H(p)$ strongly separating $(X \cap V)$ and Y exists (see Berge, 1959).

Let $\mu := \min\{px \mid x \in X \cap V\}$ and $\lambda := \sup\{px \mid x \in Y\}$; then $H(p; \mu)$ supports $(X \cap V)$ and $H(p; \lambda)$ supports Y. As $H(p)$ strongly separates $(X \cap V)$ and Y, $\lambda < 1 < \mu$ and $M_-(p) \supset M_-(p/\lambda) \supseteq Y$. Therefore a set

Fig. 9.4.4. The existence of a strongly separating hyperplane

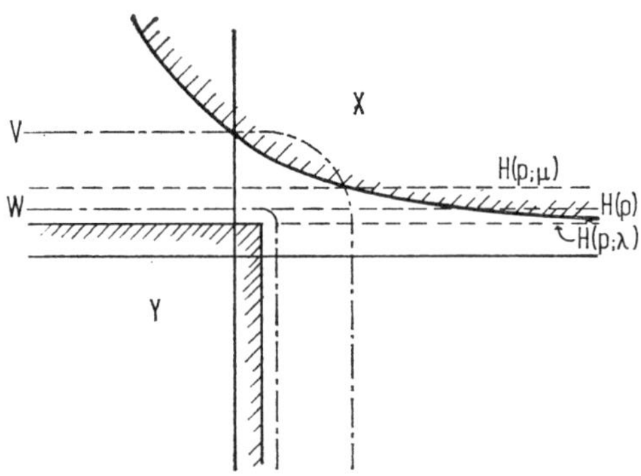

$W := B(Y; \varepsilon) \subset V$ exists such that $M_-(p) \supseteq W$ and $(X \cap V) W \cap = \phi$. Since $W \subseteq V$, we have $W \cap V = W$, from which it follows that $X \cap W = \phi$.

Because there exists a hyperplane separating X and W and $Y \subset W$, there exists a hyperplane strongly separating X and Y.

The statements (1) and (2) are thus equivalent to (1) and (2) of property 9.4.3.

When both X and Y have dimension n, then $\text{Rint } X = \text{Int } X$ and $\text{Int } X \cap \text{Int } Y = \text{Int}(X \cap Y)$. Finally, if both X and Y do not contain lines, then X^*_+ and Y^*_- also have dimension n. Statements (3) and (4) follow (see fig. 9.4.3). □

These results (especially property 9.4.4.3) may be compared with the following property which is valid for any set X and Y in R^n.

Property 9.4.5.

Let X and Y be sets in R^n. Then

$$\text{Int}(X \cap Y) \neq \phi \Rightarrow X^*_+ \cap Y^*_- = \phi.$$

Proof
Suppose that a p exists such that for all $x \in X: px \geqq 1$ and for all $y \in Y$: $py \leqq 1$. Choose $z \in \text{Int } X \cap \text{Int } Y$; then, according to properties 9.2.2.1 and 9.2.3.1., $pz > 1$ and $pz < 1$. □

It should be noticed that in the properties in this section the sets X and Y can always be substituted by the sets X^*_+ and Y^*_-.

10. Multifunctions and duality

The concept of a multifunction has been treated in 8.1, and some important topological properties of multifunctions were defined and stated in 8.2. In this chapter properties of multifunctions which depend on the linear or algebraic structure of the spaces in which and into which the multifunctions are defined will be studied. These spaces are assumed to be finite real Eucledian spaces.

Emphasis is placed on a special class of multifunctions, viz.: convex processes (section 10.3), and its subclass, the convex cone-interior processes (section 10.4).

Properties of these processes and of operations on these processes are deduced in sections 10.3 and 10.4; properties of two important cases of convex cone-interior processes are derived in section 10.5.

Firstly, however, operations on and some relevant properties of arbitrary multifunctions will be defined and stated (sections 10.1 and 10.2).

10.1. OPERATIONS ON MULTIFUNCTIONS

The algebraic operations of combinations on multifunctions are given by the following definitions. Let f and g be multifunctions from R^m into R^n, and let x be any element from $D(f) \cap D(g)$, $x_1 \in D(f)$ and $x_2 \in D(g)$. Then: *addition* of f and g is defined by:

$$(f+g)(x) := f(x)+g(x);$$

inverse addition of f and g is defined by:

$$(f \# g)(x) := \bigcup_{x_1, x_2} \{f(x_1) \cap g(x_2) \mid x_1 + x_2 = x\};$$

conjunction of f and g is defined by:

$$(f \wedge g)(x) := f(x) \cap g(x);$$

disjunction of f and g is defined by:

$$(f \vee g)(x) := \bigcup_{x_1, x_2} \{f(x_1) + g(x_2) \mid x_1 + x_2 = x\}.$$

Let f be a multifunction from $R^m \to R^n$ and λ be any scalar, then *scalar multiplication* of f by λ is defined by:

$$(\lambda f)(x) := \lambda f(x) = \{\lambda y \mid y \in f(x)\}.$$

Let $f: R^l \to R^m$ and $g: R^m \to R^n$ be multifunctions, then *multiplication* (or *composition*) of f and g is defined by:

$$(gf)(x) := g[f(x)] = \bigcup_{y} \{g(y) \mid y \in f(x)\}.$$

These operations are called *linear functional operations* and have been defined by Rockafellar (1967 and 1970). They are closely related to the operations on sets defined in 8.4 (see e.g. property 10.1.1) and can sometimes be ordered (see property 10.2.7). Examples are given in fig. 10.1.1 and 10.1.2; the operation of addition implies vertical adding, while inverse addition implies horizontal adding of the multifunctions since the graphs $G(f)$ and $G(g)$ are partially added, according to the operation # in fig. 8.3.4. Conjunction and disjunction imply intersection and addition of the graphs:

$$G(f \wedge g) = G(f) \cap G(g);$$
$$G(f \vee g) = G(f) + G(g).$$

Fig. 10.1.1. Addition (+) and inverse addition (#) of multifunctions

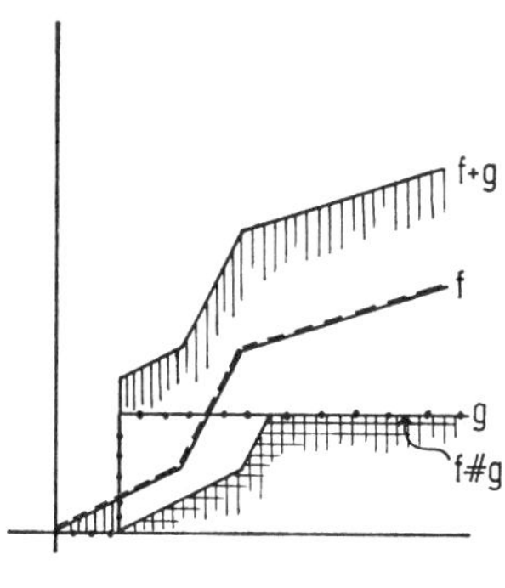

Fig. 10.1.2. Disjunction (∨) and conjunction (∧) of multifunctions

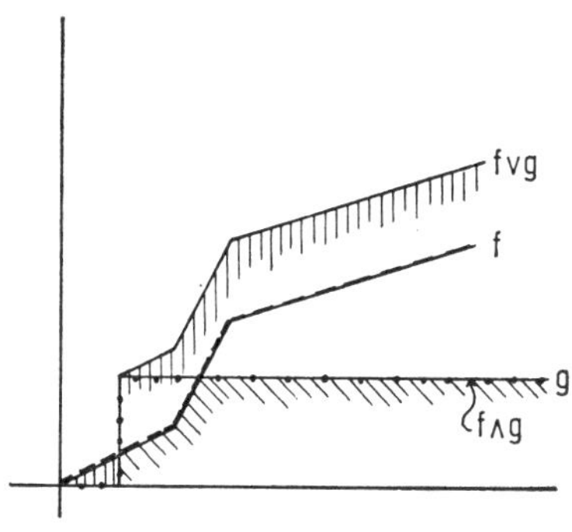

The four combinatorial operations on multifunctions can thus equivalently be defined through different modes of partial addition of the graphs of the multifunctions. A relation between the combinatorial operations on multifunctions and on sets is given by the following property:

Property 10.1.1. (operations on sets, resp. multifunctions)

Let X and Y be sets in R^n and let f and g be multifunctions from R_+ into R^n defined by $f(\lambda) := \lambda X$ and $g(\mu) := \mu Y$. Then:

$$(f + g)(1) = X + Y;$$
$$(f \# g)(1) = X \mathring{\cap} Y;$$
$$(f \wedge g)(1) = X \cap Y;$$
$$(f \vee g)(1) = X \mathring{+} Y.$$

Analogously to the convex operations defined on convex sets in section 8.3, convex operations on multifunctions can be defined on convex multifunctions. A multifunction $f: R^m \to R^n$ is said to be *convex* if its graph is a convex set.

Let f and g be convex multifunctions from R^m into R^n; let x be any element from $D(f) \cap D(g)$, $x_1 \in D(f)$, $x_2 \in D(g)$, and $0^+ \leqq \lambda \leqq 1$. Then

convex addition of f and g is defined by:

$$(f \mathring{+} g)(x) := \bigcup_{\lambda, x_1, x_2} \{(1-\lambda) f(x_1) + \lambda g(x_2) \mid x = (1-\lambda) x_1 = \lambda x_2\};$$

convex inverse addition of f and g is defined by:

$$(f \mathring{\#} g)(x) := \bigcup_{\lambda, x_1, x_2} \{(1-\lambda) f(x_1) \cap \lambda g(x_2) \mid x = (1-\lambda) x_1 + \lambda x_2\};$$

convex conjunction of f and g is defined by:

$$(f \mathring{\wedge} g)(x) := \bigcup_{\lambda, x_1, x_2} \{(1-\lambda) f(x_1) \cap \lambda g(x_2) \mid x = (1-\lambda) x_1 = \lambda x_2\};$$

convex disjunction of f and g is defined by:

$$(f \mathring{\vee} g)(x) := \bigcup_{\lambda, x_1, x_2} \{(1-\lambda) f(x_1) + \lambda g(x_2) \mid x = (1-\lambda) x_1 + \lambda x_2\}.$$

These operations are called *convex (functional) operations*. The relation between these convex functional operations and convex operations on sets may be deduced from the following equations:

$$G(f \mathring{\wedge} g) = G(f) \mathring{\cap} G(g);$$
$$G(f \mathring{\vee} g) = G(f) \mathring{+} G(g).$$

If one defines an operation for convex sets analogous to partial addition of sets, say convex partial addition, then convex conjunction and convex disjunction may be considered as being extreme cases of convex partial addition, and both convex addition and convex inverse addition as intermediate cases.

These convex functional operations will be shown to play the same role in duality theory as do the convex operations on sets (see properties 9.3.4 and 10.4.6). Convex functional operations also coincide with linear functional operations if they are applied to convex processes the graph of which is a cone:

Property 10.1.2. (*convex and linear functional operations*)

Let f and g be convex multifunctions from R^m into R^n, such that both $G(f)$ and $G(g)$ are cones containing the origin. Then:

$$f \overset{\circ}{+} g = f + g;$$
$$f \overset{\circ}{\#} g = f \# g;$$
$$f \overset{\circ}{\wedge} g = f \wedge g;$$
$$f \overset{\circ}{\vee} g = f \vee g.$$

Proof

Since $G(f)$ and $G(g)$ are convex cones, they are closed under the operations of addition and non-negative scalar multiplication (see section 8.3). Therefore, if $(x, y) \in G(f)$ and $\lambda \geqq 0$, then $(\lambda x, \lambda y) \in G(f)$, or $f(\lambda x) = \lambda f(x)$. Since in the convex operations $x_1 = x/(1-\lambda)$ and $x_2 = x/(\lambda)$, and since the argument above implies that $(1-\lambda) f(x_1) = f(x)$ and $\lambda g(x_2) = g(x)$, the definitions of the convex operations coincide with those of the linear operations. □

I turn next to dual operations on a multifunction. These operations can be defined pointwise and on the graph of the multifunction. Pointwise dual operations are called *polarity operations*, as the polar set is defined for each image set (see section 9.2).

Let $f\colon R^m \to R^n$ be a multifunction. Then
the *lower polar multifunction* $\underset{-}{*}f\colon R^m \to R^{n*}$ is defined by

$$\underset{-}{*}f(x) := [f(x)]^*_- , \qquad \text{for all } x \in D(f);$$

the *upper polar multifunction* $\underset{+}{*}f\colon R^m \to R^{n*}$ is defined by

$$\underset{+}{*}f(x) := [f(x)]^*_+ , \qquad \text{for all } x \in D(f).$$

Property 10.1.3. (continuity of polar multifunctions)

Let $f: R^m \to R^n$ be a point-closed and point-convex multifunction such that f is either point-starred or point-aureoled; let $^*f$ be correspondingly $\overset{*}{_-}f$ and $\overset{*}{_+}f$. Then:
1. f being u.h.c. implies that $^*f$ is l.h.c;
2. f being l.h.c. implies that $^*f$ is graph-closed.

Proof

1. Let f be point-starred and u.h.c. (by property 8.2.2. f is also graph-closed).

Fig. 10.1.3. f point-starred and u.h.c.

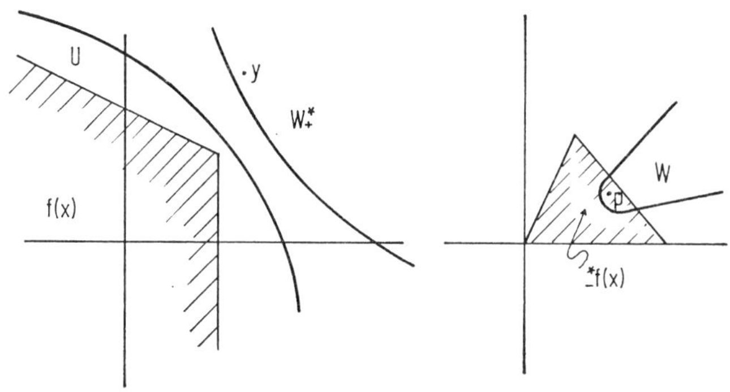

Choose $x \in D(f)$, $p \in \overset{*}{_-}f(x)$ such that $p \neq 0$, and an open set W containing p (see fig. 10.1.3). Since $\overset{*}{_-}f$ is point-starred and point-convex, W may be chosen aureoled without loss of generality ($\forall z \in D(f): f(z) \cap W = \phi \Rightarrow f(z) \cap \text{Aur } W = \phi$).

From property 9.4.2.1 it follows that $f(x)$ and W^*_+ are strongly separated. Choose an open set $U \supseteq f(x)$; since f is u.h.c., an open set $V \subseteq D(f)$ exists such that for all $z \in V$, $f(z) \subseteq U$, or $f(z)$ and W^*_+ are strongly separated (see property 8.2.1.1). Again from property 9.4.2.1, it follows that for all $z \in V$, Int $(\overset{*}{_-}f(z) \cap \text{Cl } W) \neq \phi$, or $\overset{*}{_-}f(z) \cap W \neq \phi$.

If $p \in W$ was chosen above such that $p = 0$, then evidently for all $z \in V$, $0 \in \overset{*}{_-}f(z) \cap W$. Therefore, by property 8.2.1.2, $\overset{*}{_-}f$ is l.h.c.
An analogous argument can be applied in the case of f being point-aureoled. For this case, Weddepohl (1973, p. 21) has shown that $\overset{*}{_+}f$ is l.h.c. if f is graph-closed and an ε exists such that for all $x \in D(\overset{*}{_-}f)$, $\overset{*}{_-}f(x) \cap B(0; \varepsilon) = \phi$.

2. Let f be point-starred and l.h.c.
Choose $x \in D(f)$, $p \notin {}^*_-f(x)$ and an open set W containing p, such that ${}^*_-f(x)$ and W are strongly separated. From property 9.4.2.1, it follows that $f(x)$ and W^*_+ have interior points in common, or $f(x) \cap \text{Int } W^*_+ \neq \phi$. Since f is l.h.c., there exists an open set $V \subseteq D(f)$ such that for all $z \in V$, $f(z) \cap \text{Int } W^*_+ \neq \phi$.

Therefore ${}^*_-f(z) \cap \text{Cl } W = \phi$ and ${}^*_-f(z) \cap W = \phi$, implying that ${}^*_-f$ is graph-closed.

The case of f being point-aureoled has been proven by Weddepohl (1973, p. 21). □

Dual operations which are defined on the graph of the multifunction are called *adjoint operations*. It will be shown in 10.3.4 that the adjoint of a multifunction is a generalization of the well-known adjoint of a linear transformation, which is defined as follows: let $f: R^m \to R^n$ be a linear transformation; then the adjoint $f^*: R^{n*} \to R^{m*}$ is defined by $qf(x) = f^*(q)x$, for all q and x.

The adjoint defined below is also a generalization of the adjoint defined by Rockafellar (1972) for convex cone processes (see section 10.3). The analogy between both definitions is of the same nature as the analogy between linear and convex operations drawn above.

Due to the fact that only in the linear case $f(-x) = -f(x)$, it is necessary to distinguish two adjoint operations:
Let $f: R^m \to R^n$ be a multifunction. Then
the *lower adjoint multifunction* $f^*_-: R^{n*} \to R^{m*}$ is defined by:

$$f^*_-(q) := \{p \mid \forall x, \forall y \in f(x): px \leqq qy + 1\};$$

the *upper adjoint multifunction* $f^*_+: R^{n*} \to R^{m*}$ is defined by

$$f^*_+(q) := \{p \mid \forall x, \forall y \in f(x): px \geqq qy - 1\}.$$

It can be easily shown that the adjoint multifunctions can equivalently be defined by means of the polar set of the graph:

$$f^*_-(q) = \{p \mid (p, -q) \in [G(f)]^*_-\};$$
$$f^*_+(q) = \{p \mid (-p, q) \in [G(f)]^*_-\}.$$

This method of notation also indicates that the difference between both adjoint multifunctions is only a question of signs and is therefore of a much simpler nature than the difference between the two polar sets or multifunctions. It should be repeated that:

$${}^*_-f(x) = \{q \mid \forall y \in f(x): qy \leqq 1\};$$
$${}^*_+f(x) = \{q \mid \forall y \in f(x): qy \geqq 1\}.$$

For two special classes of multifunctions, the relation between the polar and adjoint multifunction is derived (see property 10.5.3).

The various dual and inverse multifunctions are related as follows:

The properties of the operations will be analyzed in section 10.3, et al.

10.2 PROPERTIES OF MULTIFUNCTIONS

Let $f: R^n \to R^n$ be a multifunction on R^n into itself; then $x \in R^n$ is called a *fixed point* of f if $x \in f(x)$. The following property has been shown by Kakutani (1941).

Property 10.2.1. (*Kakutani fixed point theorem*)

Let X be a non-empty, compact and convex subset of R^n; if f is a point closed, point convex, u.h.c. multifunction from X into X such that $D(f) = X$, then f has a fixed point.

Fig. 10.2.1. Fixed points of a multifunction $f: X \to X$

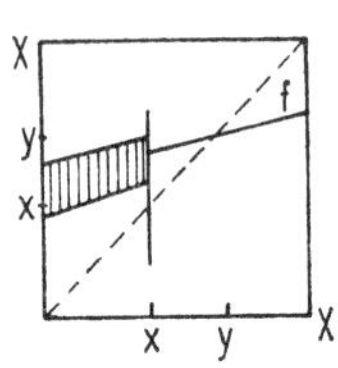

This theorem has been extended by Ky Fan (1952) to topological spaces other than the Euclidean n-space. The Kakutani fixed point theorem is a generalization of a fixed point theorem for functions which has been proven by Brouwer (1909):

Property 10.2.2. (*Brouwer fixed point theorem*)

Let X be a non-empty, compact and convex subset of R^n; if f is a continuous function from X into X, then f has a fixed point, i.e. an $x \in X$ exists such that $x = f(x)$.

A *multifunction* $f: R^m \to R^n$ is said to be *convex* if its graph is a convex set. A multifunction $f: R^m \to R^m$ with a convex effective domain is convex if and only if for all x and y in $D(f)$ and $\lambda \in [0, 1]$,

$$[(1-\lambda) f(x) + \lambda f(y)] \subseteq f((1-\lambda)x + \lambda y).$$

A multifunction $f: R^m \to R^n$ is said to be *quasi-convex* if its inverse multifunction $f^{-1}: R^n \to R^m$ is point-convex. A multifunction $f: R^m \to R^n$ with a convex effective domain is quasi-convex if and only if for all x and y in $D(f)$ and $\lambda \in [0, 1]$,

$$[f(x) \cap f(y)] \subseteq f((1-\lambda)x + \lambda y).$$

Both properties have been shown by Jacobson (1970); quasi-convex multifunctions are represented in fig. 10.1.1.

Let $f: D \to R^1$ be a real valued *function* with $D \subseteq R^m$. The function f is said to be *concave* if its less closure $f_-: D \to R^1$ is a convex multifunction, i.e. if $G(f_-)$ is a convex set. The function f is concave if and only if for all x and y in D and $\lambda \in [0, 1]$,

$$[(1-\lambda) f(x) + \lambda f(y)] \leqq f((1-\lambda)x + \lambda y).$$

The function f is said to be *convex* if $-f$ is concave.

The function f is said to be *quasi-concave* if its less closure f_- is a quasi-convex multifunction, i.e. if $(f_-)^{-1}$ is point-convex. The function f is quasi-concave if and only if for all x and y in D and $\lambda \in [0, 1]$,

$$\min\{f(x), f(y)\} \leqq f((1-\lambda)x + \lambda y).$$

The function f is said to be *quasi-convex* if $-f$ is quasi-concave.
Let f and g be real valued functions. If f and g are quasi-concave, the algebraic operations are defined as follows (compare figs. 10.1.1 and 10.1.2):

$$(f+g)(x) := f(x) + g(x);$$
$$(f \# g)(x) := \sup\{\min\{f(x_1), g(x_2)\} \mid x = x_1 + x_2\};$$
$$(f \wedge g)(x) := \min\{f(x), g(x)\};$$
$$(f \vee g)(x) := \sup\{f(x_1) + f(x_2) \mid x = x_1 + x_2\}.$$

A multifunction $f: R^m \to R^n$ is said to be *quasi-concave* if its inverse multifunction is point-concave, i.e. if $D(f) \backslash f^{-1}(y)$ is convex for all $y \in R(f)$.

An example of a quasi-concave multifunction is the preference multifunction C defined in section 3.8.

It can be shown that a multifunction $f\colon R^m \to R^n$ with a convex effective domain is quasi-concave if and only if for all x and y in $D(f)$ and λ in $[0, 1]$:

$$[f(x) \cup f(y)] \supseteq f((1-\lambda)x+\lambda y).$$

Choosing x, y from $D(f)\backslash f^{-1}(z)$ implies that $z \notin f(x) \cup f(y)$, or $[(1-\lambda)x+\lambda y] \in D(f)\backslash f^{-1}(z)$, for any $\lambda \in [0, 1]$; therefore f^{-1} is point-concave. Conversely, suppose that for some λ, there exists a $z \in f((1-\lambda)x+\lambda y)$ and $z \notin f(x) \cup f(y)$. Then $[(1-\lambda)x+\lambda y] \notin D(f)\backslash f^{-1}(z)$ and $x, y \in D(f)\backslash f^{-1}(z)$. This contradicts the convexity of $D(f)\backslash f^{-1}(z)$.

Comparison with the operations on multifunctions (defined in section 10.1) makes it clear that for quasi-concave functions max (or sup) is substituted for union, and min (or inf) is substituted for intersection. In most cases, the inclusion sign $\subseteq$ used for multifunctions replaces the inequality sign $\leqq$ used for functions. An exception is formed by the (quasi-) convex (resp. concave) function defined above, for which the inequality sign is the opposite of the inclusion sign for (quasi-) convex (resp. concave) multifunctions. The definition of a concave function, however, in too widely accepted for me to change its name here into a convex function. The definitions in this book follow the definition that a convex multifunction has a convex graph.

Let $f\colon R^m \to R^n$ be a multifunction. Then f is said to be *starred* if f is point-starred and f^{-1} is point-aureoled; f is said to be *aureoled* if f^{-1} is starred.

These definitions imply that $D(f)$ is an aureoled set whenever f is starred (see fig. 10.2.3) and that $R(f)$ is a starred set if f is aureoled (see fig. 10.2.2).

Let $f\colon R^m \to R^n$ be a multifunction. Then f is said to be *homothetic* if for all $x_1, x_2 \in D(f)$, a $\mu > 0$ exists such that $f(x_1) = \mu f(x_2)$.

The following property follows from property 9.3.4.5:

Property 10.2.3. (*homotheticity and duality*)

A starred (resp. aureoled) multifunction $f\colon R^m \to R^n$ is homothetic if and only if its polar multifunction $^*f\colon R^m \to R^{n*}$ is homothetic.

If, for example, for some $x_1, x_2 \in D(f)\colon f(x_1) = \mu f(x_2)$, then $\mu\, {}^*f(x_1) = {}^*f(x_2)$. It should be noticed that the union of the boundaries of the image sets does not need to be equal to $R(f)$, as is shown in fig. 10.2.2.

Fig. 10.2.2. An aureoled homothetic multifunction, $f: R_+ \to R_+^2$, and the boundaries of its image sets in R^2.

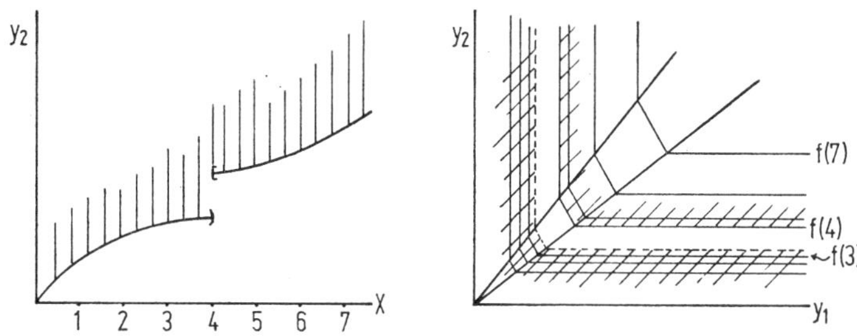

Let $f: R^m \to R^n$ be a multifunction such that $D(f)$ is a cone. Then f is said to be *homogeneous of degree* k if for all non-zero $x \in D(f)$ and $\lambda > 0$:

$$f(\lambda x) = \lambda^k f(x);$$

f is said to be *linear homogeneous* if f is homogeneous of degree 1.

Property 10.2.4. (*homogeneity of inverses and polars*)

Let $f: R^m \to R^n$ be a multifunction.

1. f is homogeneous of degree k if and only if f^{-1} is homogeneous of degree $1/k$;
2. If f is starred (resp. aureoled) then: f is homogeneous of degree k if and only if $^*f$ is homogeneous of degree $-k$.

Proof

1. $G(f) = \{(\lambda x, \lambda^k y) \mid y \in f(x)\} = \{(\lambda^l x, \lambda y) \mid y \in f(x),\ l := 1/k\}$, as by definition f is homogeneous; from:

$$G(f^{-1}) = \{(y, x) \mid (x, y) \in G(f)\} = \{(\lambda y, \lambda^l x) \mid x \in f^{-1}(y),\ l = 1/k\}$$

the statement follows.

2. From property 9.3.4.5 it follows that $[\lambda^k f(x)]^* = \lambda^{-k}[f(x)]^*$;
the statement follows from the definition of $^*f$ (see also fig. 10.2.3 and property 10.5.1.5). □

Fig. 10.2.3. A starred homogeneous multifunction of degree 2, and its polar multifunction, homogeneous of degree −2.

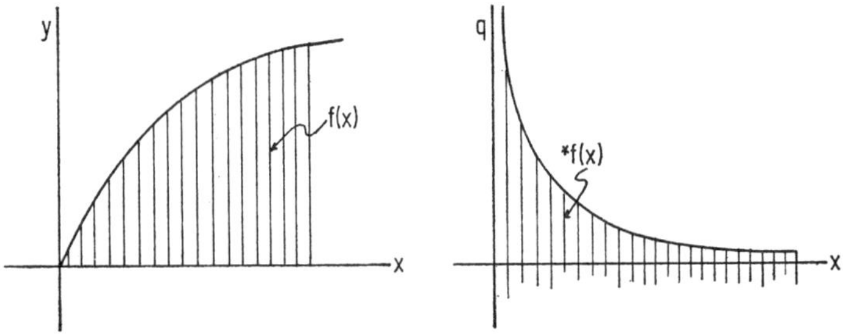

Let $f: R^m \to R^n$ be a multifunction and let x and y be elements of $D(f)$. Then f is said to be *super additive* if

$$f(x)+f(y) \subseteq f(x+y);$$

f is said to be *sub additive* if

$$f(x)+f(y) \supseteq f(x+y).$$

As a convex cone is closed under the operations of addition and non-negative scalar multiplication, the condition that a multifunction be super additive and linear homogeneous is equivalent to the graph of the multifunction being a convex cone.

Let $f: R^m \to R^n$ be a multifunction and let x_1 and x_2 be any pair of elements of $D(f)$, and y_1, y_2 any pair of elements of $R(f)$. If

$x_1 \leqq x_2$ implies $f(x_1) \subseteq f(x_2)$

and $y_1 \leqq y_2$ implies $f^{-1}(y_1) \supseteq f^{-1}(y_2)$, then f is said to be a *monotone increasing* multifunction. A multifunction is said to be *monotone decreasing* if its inverse is a monotone increasing multifunction.

Monotone multifunctions are characterized by the fact that the image sets contain all elements greater (resp. smaller) than any given element in the set according to the partial order relation $\leqq$ (compare fig. 10.2.4 and 10.2.5).

Property 10.2.5. (monotone multifunctions)

Let $f: R^m \to R^n$ be a multifunction. Then the following statements under 1, resp. 2, are equivalent:

1. a. f is monotone increasing;
   b. $f(\text{Less } x) = f_-(x) = f(x)$;
   c. $\text{Less}\{(-x, y) \mid y \in f(x)\} = \{(-x, y) \mid y \in f(x)\}$.
2. a. f is monotone decreasing;
   b. $f(\text{More } x) = f_+(x) = f(x)$;
   c. $\text{More}\{(-x, y) \mid y \in f(x)\} = \{(-x, y) \mid y \in f(x)\}$.

Proof

1. b. $f(\text{Less } x) = \cup\{f(z) \mid z \leqq x\} = \cup\{f(z) \mid f(z) \subseteq f(x)\} = f(x)$.
$y_1 \leqq y_2 \Rightarrow f^{-1}(y_1) \supseteq f^{-1}(y_2)$ is equivalent to
$y_1 \leqq y_2 \in f(x) \Rightarrow y_1 \in f(x)$, or $f_-(x) \subseteq f(x)$;
since f_- is a closure of f, $f_- \supseteq f$; therefore $f_- = f$.
1. c. Choose any $(-\bar{x}, \bar{y}) \in \text{Less}\{(-x, y) \mid y \in f(x)\} = \{(-\tilde{x}, \tilde{y}) \mid \exists y \in f(x) : -\tilde{x} \leqq -x \text{ and } \tilde{y} \leqq y\}$.
Since $-\bar{x} \leqq -x$, or $\bar{x} \geqq x$ implies $f(\bar{x}) \supseteq f(x)$ and as $\bar{y} \leqq y \in f(x)$ implies $\bar{y} \in f(x)$, it follows that $(-\bar{x}, \bar{y}) \in \{(-x, y) \mid y \in f(x)\}$.

The inverse inclusion, $\supseteq$, is implied by the fact that a less closure always contains the closed set.

2. Similar arguments can be applied to monotone decreasing multifunctions (see fig. 10.2.5). □

Fig. 10.2.4. A monotone increasing multifunction

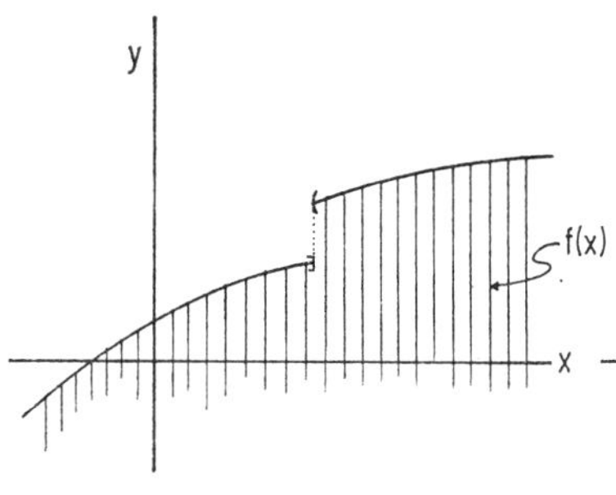

Fig. 10.2.5. A monotone decreasing multifunction

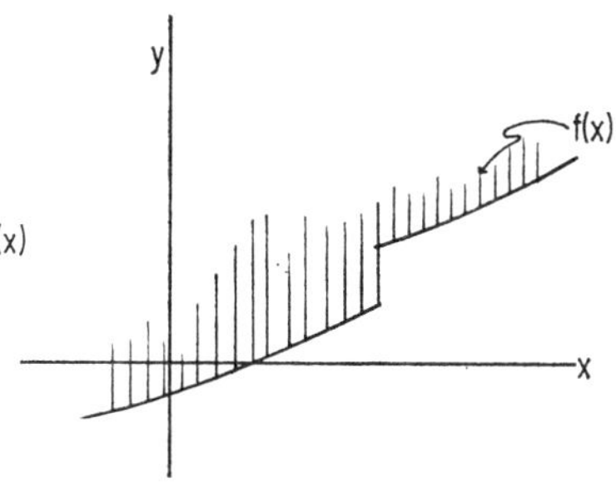

A multifunction $f: R^m_+ \to R^n_+$ is said to be a *non-negative monotone multifunction* if f is monotone increasing or decreasing, $D(f)$ is a cone in R^m_+ and $R(f)$ is a cone in R^n_+. Non-negative monotone multifunctions play an important part in economic theory (see chapters 3 and 4).

Property 10.2.6. (*non-negative monotone multifunctions*)

Let $f: R^m_+ \to R^n_+$ be a non-negative monotone multifunction.
If f is monotone increasing, then f is starred.
If f is monotone decreasing, then f is aureoled.

Proof
Let f be monotone increasing; then for all $y_1, y_2 \in R(f)$, $y_1 \leqq y_2$ implies $f^{-1}(y_1) \supseteq f^{-1}(y_2)$. Choose $x \in f^{-1}(y_2)$, then $0 \leqq y_1 \leqq y_2$ implies $y_1 \in f(x)$, or $[0, y_2] \subseteq f(x)$. Thus $f(x)$ is point starred. Conversely, choose some $\bar{x} \in D(f)$ and some $\bar{y} \in f(\bar{x})$. If $f^{-1}(\bar{y})$ is not an aureoled set, then a $\lambda > 1$ exists such that $\lambda\bar{x}$ is not an element of $f^{-1}(\bar{y})$, or $\bar{y} \notin f(\lambda\bar{x})$. As $0 \leqq \bar{x}$ and $\bar{x} < \lambda\bar{x}$, it follows by definition that $\bar{y} \in f(\bar{x}) \subseteq f(\lambda\bar{x})$. This contradicts $f^{-1}(\bar{y})$ not being aureoled.
A similar reasoning can be applied to show the second implication. □

The algebraic operations of combinations can be rather simply ordered if the multifunctions are non-negative monotone. The following property can be easily verified from the definitions (see also figs. 10.1.1 and 10.1.2):

Property 10.2.7. (*algebraic operations on monotone multifunctions*)

Let f and g be non-negative monotone multifunctions from R^m_+ into R^n_+.
If they are monotone increasing, then:

$$(f \# g) \subseteq (f \wedge g) \subseteq (f \vee g) \subseteq (f+g);$$

if they are monotone decreasing, then:

$$(f \# g) \supseteq (f \wedge g) \supseteq (f \vee g) \supseteq (f+g).$$

10.3. CONVEX PROCESSES

Let $f: R^m \to R^n$ be a multifunction; f is said to be a *convex process* if its graph, $G(f)$, is a closed and convex set (see fig. 10.3.1). Any convex process $f: R^m \to R^n$ determines uniquely a closed and convex set, i.e. its graph, in R^{m+n}.
On the other hand, it is possible to determine a unique convex *process* from a given closed and convex *set* if a bisection of the Euclidean space in which the set is contained is also given. A *bisection* of a real Euclidean n-space is said to be an ordered pair of non-zero subspaces (R^l, R^m) such that their direct sum is equal to the real Euclidean n-space, i.e. $R^l \oplus R^m = R^n$.

Fig. 10.3.1. A convex process f and its inverse f^{-1}

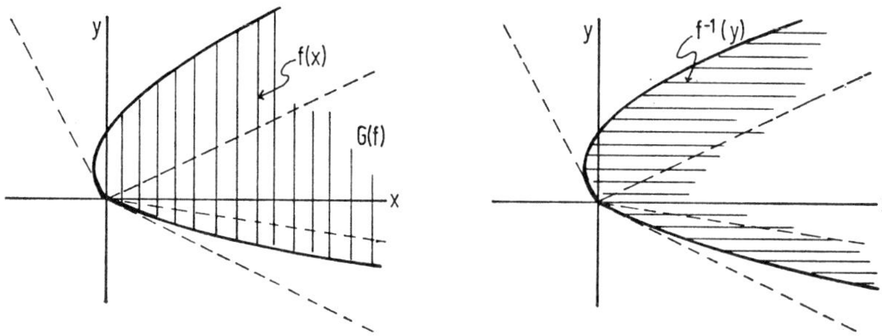

If, for example, the closed and convex set $F := G(f)$ in fig. 10.3.1 is given and R^2 is bisected in (X, Y), resp. the x-axis and the y-axis, then the convex process $f: X \to Y$ is defined by $f(x) := \{y \mid (x, y) \in F\}$. The bisection of R^2 in (Y, X) generates the convex process $g(y) := \{x \mid (x, y) \in F\}$, which is equal to the inverse of f, i.e. $g = f^{-1}$.

Property 10.3.1. (*convex processes*)

Let $f: R^m \to R^n$ be a convex process. Then:
1. for all $x, y \in D(f)$ and $\lambda \in [0, 1]$:
   $(1-\lambda) f(x) + \lambda f(y) \subseteq f((1-\lambda)x + \lambda y)$;
2. f is point-convex;
3. f is point-closed;
4. f is lower hemi-continuous;
5. f^{-1} is a convex process;
6. f_+^* and f_-^* are convex processes.

Proof
1. Equivalent to f being convex (see 10.2).

2. The intersection of two convex sets is convex (see 8.3.3).

3. The intersection of two closed sets is closed (see 8.2).

4. Let W be an open convex set in $R(f)$.
If $V := f^{-1}(W) = \{x \mid f(x) \cap W \neq \phi\}$ is open in $D(f)$, then f is l.h.c. Assume that V is not open (see fig. 10.3.2): then there exists an $\bar{x} \in \text{Bnd } V \cap V$. Since f is a convex process, V is a convex set and an open convex set M exists such that $\bar{x} \in \text{Cl } M$, and for each $x \in M: f(x) \cap W = \phi$ (otherwise $x \in V$ and V would be open).

Fig. 10.3.2. A multifunction which is not l.h.c.

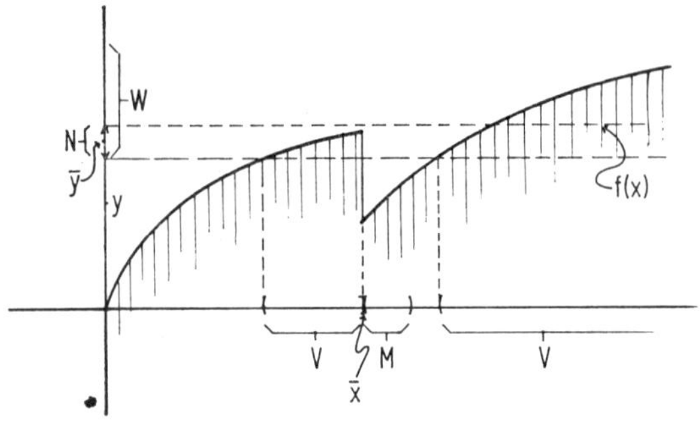

For some $\delta>0$ and $\bar{y}\in W\cap f(\bar{x})$, the set $N:=\{y\in W\,|\,|y-\bar{y}|<\delta\}$ is an open and convex set such that for each $x\in M: f(x)\cap N=\phi$. Choose any $z\in M$ and $y\in f(z)$; then a $\lambda\in(0, 1)$ exists such that $\lambda|y-\bar{y}|<\delta$, or $\bar{y}+\lambda(y-\bar{y})\in N$. Since M is convex, $\bar{x}+\lambda(z-\bar{x})\in M$; this implies that $f(\bar{x}+\lambda(z-\bar{x}))\cap N=\phi$, thus contradicting the convexity of f, by which $f(\bar{x}+\lambda(z-\bar{x}))\supseteq f(\bar{x})+\lambda(f(z)-f(\bar{x}))$, which latter set contains $\bar{y}+\lambda(y-\bar{y})\in N$.

5. $G(f^{-1})=\{(y, x)\,|\,y\in f(x)\}$ is a closed and convex set.

6. $G(f^*_-)=\{(q, p)\,|\,(p, -q)\in[G(f)]^*_-\}$ is closed and convex (from property 9.2.3.4). The same is true for $G(f^*_+)$ (see property 9.2.2.4). Compare also fig. 10.3.3. □

It must be emphasized that a convex process need not be upper semi continuous. This can be verified from the following counter example:

Consider the function $\bar{f}: R^2_+\to R_+$ defined by

$$x=\bar{f}(l, k):=(l^{-2}+k^{-2})^{-0,5}.$$

This function belongs to the class of constant elasticity of substitution-production functions developed by Arrow, Chenery, Minhas and Solow (1961). The function $\bar{f}$ is linear homogeneous and concave. The non-negative less closure of $\bar{f}$ is the multifunction $f: R^2_+\to R_+$ defined by

$$f(l, k):=\{x\,|\,x\leqq(l^{-2}+k^{-2})^{-0,5}\}.$$

The inverse of f is the multifunction $g: R_+\to R^2_+$ defined by

$$g(x):=\{(l, k)\,|\,x\leqq(l^{-2}+k^{-2})^{-0,5}\}=f^{-1}(x).$$

As $G(g)$ is closed and convex, g is a convex process; in fact, g is a convex cone process.

Fig. 10.3.3. Duality operations related to f

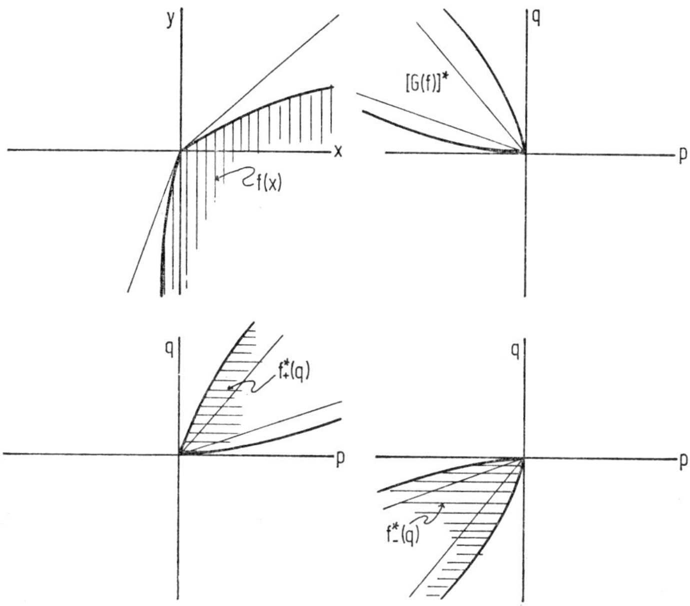

Next, take a closed set $A := \{(l, k) \mid 0 \leqq l,\ 0 \leqq k \leqq 1\}$ in $R(g)$, as shown in fig. 10.3.4. The image of A is equal to $g^{-1}(A) = f(A) = \{x \in R_+ \mid x \leqq (l^{-2}+1)^{-0,5}\} = [0, 1)$. Since $g^{-1}(A)$ is not closed, g is not u.h.c.

Fig. 10.3.4. A convex process which is not u.h.c.

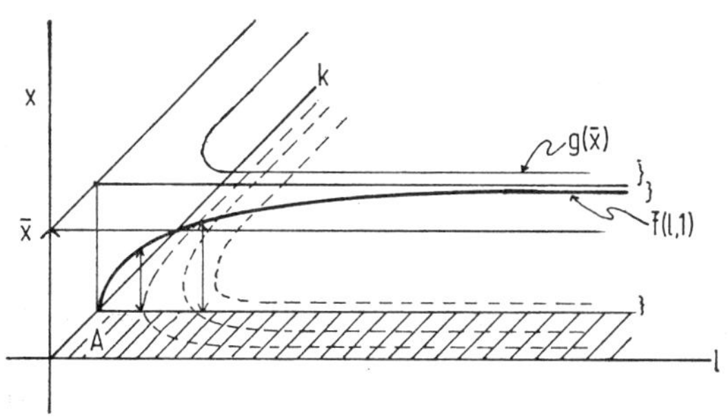

Let $f: R^m \to R^n$ be a convex process: then f is said to be a *convex cone process* if $G(f)$ is a cone (see fig. 10.3.5); f is said to be a *convex polyhedral*

process if $G(f)$ is a polyhedral cone, i.e. if $G(f)$ can be expressed as the intersection of a finite collection of closed half-spaces.

Property 10.3.2. (*convex cone processes*)

Let $f: R^m \to R^n$ be a convex cone process. Then:

1. f is superadditive, i.e. for any $x, y \in D(f)$:
   $f(x+y) \supseteq f(x) + f(y)$;
2. f is linear homogeneous, i.e. for any $x \in D(f)$ and $\lambda > 0$,
   $f(\lambda x) = \lambda f(x)$;
3. $0 \in f(0)$;
4. f^{-1} is a convex cone process;
5. f_-^* and f_+^* are convex cone processes;
6. $f(0)$ is the recession cone of $f(x)$ for all $x \in D(f)$.

Proof

Conditions (1), (2) and (3) are necessary and sufficient for $G(f)$ to be a convex cone (see 8.3); e.g. $(a, x) \in G(f)$ and $(b, y) \in G(f)$ imply that $(a+b, x+y) \in G(f)$. Condition (4) follows from $G(f^{-1}) = \{(y, x) \mid (x, y) \in G(f)\}$. Condition (5) follows from property 9.2.5.3, implying that $[G(f)]_-^* = [G(f)]_-^0$, which is a closed convex cone.

Condition (6) follows from property 8.3.1 which states that Conint $G(f)$ equals the recession cone. As Conint $G(f) = G(f)$, $f(0) = G(f) \cap \{(0, y) \mid y \in R^n\}$ determines the set of directions parallel to R^n in which $G(f)$ is unbounded. Thus $f(0)$ is the recession cone of $f(x) = \{y \mid (x, y) \in G(f)\}$ for every $x \in D(f)$. □

The two following convex cone processes can be associated with any convex process $f: R^m \to R^n$. The *cone closure* of f is said to be the multifunction $f_c: R^m \to R^n$ defined by

$$f_c(x) := \{y \mid (x, y) \in \text{Cone } G(f)\};$$

the *cone opening* of f is said to be the multifunction $f_o: R^m \to R^n$ defined by

$$f_o(x) := \{y \mid (x, y) \in \text{Conint } G(f)\}.$$

As both Cone $G(f)$ and Conint $G(f)$, if non-empty, are closed and convex cones with vertex 0, the multifunctions f_c and f_o are convex cone processes (see fig. 10.3.5).

Convex polyhedral processes belong to a class of convex cone processes for which many important properties have been derived. A well known example is the *von Neuman process* $f: R^n \to R^n$ defined for two matrices A and B by $f(x) := \{y \mid \exists z \in R_+^m: 0 \leqq x \leqq Az \text{ and } 0 \leqq y \leqq Bz\}$.

Fig. 10.3.5. Convex cone processes: the cone closure f_c and the cone opening f_o of the process f in fig. 10.3.1.

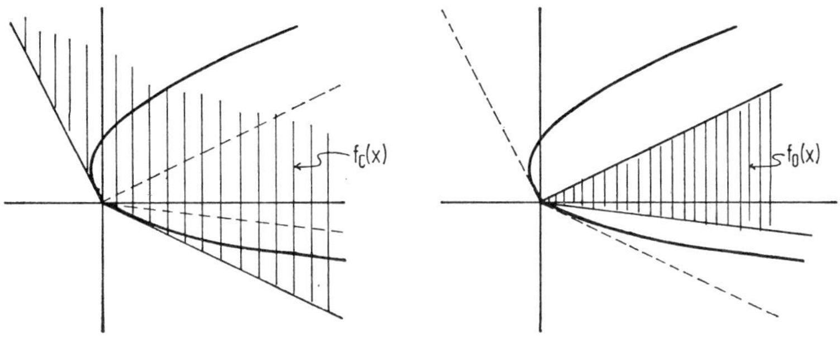

These convex polyhedral processes have been studied by Rockafellar (1972) who has proven the following property:

Property 10.3.3. (*convex polyhedral processes*)

Let f and g be two convex polyhedral processes from R^m into R^n, and h a convex polyhedral process from R^l into R^m:

1. Then the processes $\lambda f, f+g, f \wedge g, f \vee g$ and fh are all convex polyhedral processes;
2. let f^* be defined by $f^*(q) := \{p \mid \forall x, \forall y \in f(x)\colon\ px \leqq qy\}$; then:

$(f+g)^* = f^* + g^*$; $\quad (fh)^* = h^* f^*$;

$(f \wedge g)^* = f^* \vee g^*$; $\quad (f \vee g)^* = f^* \wedge g^*$;

$(\lambda f)^* = \lambda f^*$, for $\lambda > 0$.

This property is not valid for arbitrary convex cone processes, because operations on those processes do not always preserve closedness of the graphs of the processes. In the next section a special class of convex processes will be defined, viz., the convex cone-interior process, which has properties analogous to those of a convex polyhedral process and which is a generalization of that process..

A multifunction $f\colon R^m \to R^n$ is said to be a *superlinear process* if it is a convex cone process such that $f(0) = \{0\}$. The theory of superlinear processes may be considered as a generalization of the theory of non-negative matrices. Spectral properties and optimal trajectories for convex polyhedral processes have been studied by Rockafellar (1972) and for superlinear processes by Makarov and Rubinov (1970). These dynamic properties will not be treated in this study.

Property 10.3.4. (*Inverse and adjoint processes*)

Let $f\colon R^m \to R^n$ be a convex process. Then:

1. $(f^*_-)^*_+ = (f^*_+)^*_- = f$;
2. $(f^*_-)^{-1} = (f^{-1})^*_+$;
   $(f^*_+)^{-1} = (f^{-1})^*_-$.

Proof
1. The graph of f^*_- is equal to $G(f^*_-) = \{(q, p) \mid (p, -q) \in [G(f)]^*_-\}$; the lower polar set of this graph equals $[G(f^*_-)]^*_- = \{(y, x) \mid (x, -y) \in G(f)\}$, as $G(f)$ is reflexive under this operation (see property 9.2.2.4). This set generates the adjoint process of $f^*_-(q)$, which by definition is equal to $(f^*_-)^*_+(x) = \{y \mid (-y, x) \in [G(f^*_-)]^*_-\}$. For any x, the set $(f^*_-)^*_+(x)$ reduces to $\{y \mid (x, y) \in G(f)\} = f(x)$.
A similar argument is valid for $(f^*_+)^*_-$.

2. By definition (see 10.1), $G(f^*_-) = \{(q, p) \mid (p, -q) \in [G(f)]^*_-\}$. The inverse $(f^*_-)^{-1}$ gives $G[(f^*_-)^{-1}] = \{(p, q) \mid (p, -q) \in [G(f)]^*_-\}$. Conversely, $G(f^{-1}) = \{(y, x) \mid (x, y) \in G(f)\}$ and $G[(f^{-1})^*_+] = \{(p, q) \mid (-q, p) \in [G(f^{-1})]^*_-\} = \{(p, q) \mid (p, -q) \in [G(f)]^*_-\} = G[(f^*_-)^{-1}]$.
A similar argument is valid for the other equation. □

Property 10.3.4 indicates in addition that both the adjoint operation and the inverse operation are 'orientation reversing'; this can also be seen from figures 10.3.1 and 10.3.3.

Property 10.3.5. (*adjoints*)

Let $f\colon R^m \to R^n$ be a *convex process* and let f^*_+ and f^*_- from R^{n*} into R^{m*} be the adjoint processes of f, i.e., for all q and x:

$$f^*_-(q) \cdot x \leqq q \cdot f(x) + 1;$$
$$f^*_+(q) \cdot x \geqq q \cdot f(x) - 1.$$

If $G(f)$ is a cone, i.e. if f is a *convex cone process*, then, for all q and x:

$$f^*_-(q) \cdot x \leqq q \cdot f(x)$$
$$f^*_+(q) \cdot x \geqq q \cdot f(x),$$

and if $G(f)$ is a subspace with $f(0)$ bounded, i.e. if f is a *linear process*, then, for all q and x:

$$f^*_-(q)\cdot x = f^*_+(q)\cdot x = q\cdot f(x).$$

Proof

According to property 9.2.5.3, $[G(f)]^0_- = [\text{Cone } G(f)]^*_-$. Therefore, if $G(f)$ is a cone then $px+qy \leqq 1$ implies $px+qy \leqq 0$, for all $(p, q)\in[G(f)]^*_-$, or $px \leqq (-q)y+1$ implies $px \leqq (-q)y$. If $G(f)$ is a subspace, then $y\in f(x)$ implies $-y\in f(-x)$. Thus $f^*(q) = \{p \mid px \leqq qy$ and $px \geqq qy$; for all $y\in f(x)$ and $x\}$. □

When f is a linear transformation, f may be represented by a matrix A; the adjoint process f^* is then equal to A^*, the transpose of matrix A, and $(A^*q)x = q(Ax)$. Therefore f^* may be considered as a generalization of the adjoint in linear algebra. This also explains the change of sign in the definitions given on page 191 of the adjoint multifunctions.

The duality theorem in (non-)linear programming can be generalized by means of the adjoint process. The elements of the adjoint process can be seen as vectors of Lagrange-multipliers corresponding to the constraints generated by a convex polyhedral process.

Let $f: R^m \to R^n$ be a starred convex process. The *primal problem* is to find a vector $y\in f(\bar{x})$ such that $\bar{q}y$ is maximized. The *dual problem* is to find a vector $p\in f^*_+(\bar{q})$ such that $p\bar{x}$ is minimized. The following property can be demonstrated:

Property 10.3.6. (duality theorem)

Let the function $\mu : f(\bar{x})\times f^*_+(\bar{q})\to R^1$, defined by $\mu(y, p) := \bar{q}y+p\bar{x}$ have a saddle point $(\bar{y}, \bar{p})$, i.e.

$$\mu(y,\bar{p}) \leqq \mu(\bar{y}, \bar{p}) \leqq \mu(\bar{y}, p).$$

Then $\bar{y}$ solves the primal problem, $\bar{p}$ solves the dual problem and $\bar{q}\bar{y} = \bar{p}\bar{x}+1$.

Proof

$\mu(y, \bar{p}) \leqq \mu(\bar{y}, \bar{p})$ is equivalent to

$\max\limits_{y\in f(x)} \mu(y, \bar{p}) = \mu(\bar{y}, \bar{p}) = \bar{q}\bar{y}+\bar{p}\bar{x}$. Since $\bar{p}\bar{x}$ is constant, $\bar{y}$ is the solution to the primal problem.

Moreover, $\mu(\bar{y}, p) \geqq \mu(\bar{y}, \bar{p})$ is equivalent to

$\min\limits_{p\in f^*_+(\bar{q})} \mu(\bar{y}, p) = \mu(\bar{y}, \bar{p})$, which implies the solution of the dual problem.

From the definition of $f^*_+(\bar{q})$ it follows that $\bar{p}\bar{x} = \bar{q}\bar{y}-1$. □

A similar property is valid when the primal problem is minimization of a linear function over an aureoled constraint set (this is the dual problem above). It is evident that $\bar{q}\bar{y} = \bar{p}\bar{x}$ if f is a convex cone process. If f is also a polyhedral process and defined by $f(\bar{x}) := \{y \in R^n_+ \mid Ay \leqq \bar{x}\}$, then the relevant dual set of constraints is equal to $f^*(\bar{q}) = \{p \in R^n_+ \mid A^*p \geqq \bar{q}\}$.

10.4. CONVEX CONE-INTERIOR PROCESSES

The class of convex processes as a whole is not closed under the algebraic operations defined in section 10.1, and this is also the case for the sub-class of convex cone processes. It has been shown, however, that the sub-class of convex polyhedral processes is closed under all algebraic operations defined in section 10.1 (see property 10.3.3). In this section a convex process will be defined which is indeed closed under all algebraic operations defined in section 10.1, but which need not be a convex cone process, i.e. linear homogeneity and super additivity are not required.

In some cases, one may convert a convex set or a convex process into a convex cone, resp. a convex cone process by adding one, resp. two dimensions to the spaces in which they are defined, but this conversion to 'linear' theory cannot be again reduced to the convex case if combinatorial operations, such as addition, and duality operations are involved. Conversion to cones may be sensible when multiplying a process with itself to study trajectories of that process, but not if aggregation of processes is also involved.

Let $f: R^m \to R^n$ be a convex process; f is said to be a *convex cone-interior process* if $R(f)$ and $R(f^{-1})$ are cones and if $f(0) = f_c(0)$ and $f^{-1}(0) = f_c^{-1}(0)$.

Since f_c is the cone closure of a convex process f (see section 10.3), $f_c(0)$ is a cone and $f(0)$ is necessarily a cone. This definition implies that the graph of a convex cone-interior process is a closed, convex and unbounded set containing the origin on its boundary; it also implies that the projection of the graph on R^m, resp. R^n, is a cone. The conditions for $f(0)$ and $f^{-1}(0)$ in the definition are necessary in order to guarantee that the range and domain of the adjoint processes are cones (see property 10.4.1.2). Examples of convex cone-interior processes are given in figs. 10.4.1 and 10.5.2, which may be compared with fig. 10.3.1, 10.3.3, or 10.4.2, not giving a convex cone-interior process.

The behavior of a convex cone-interior process f in the immediate neighborhood of the origin is characterized by f_c, the cone closure of f.

Fig. 10.4.1. A convex cone-interior process f and its adjoints f_+^* and f_-^*

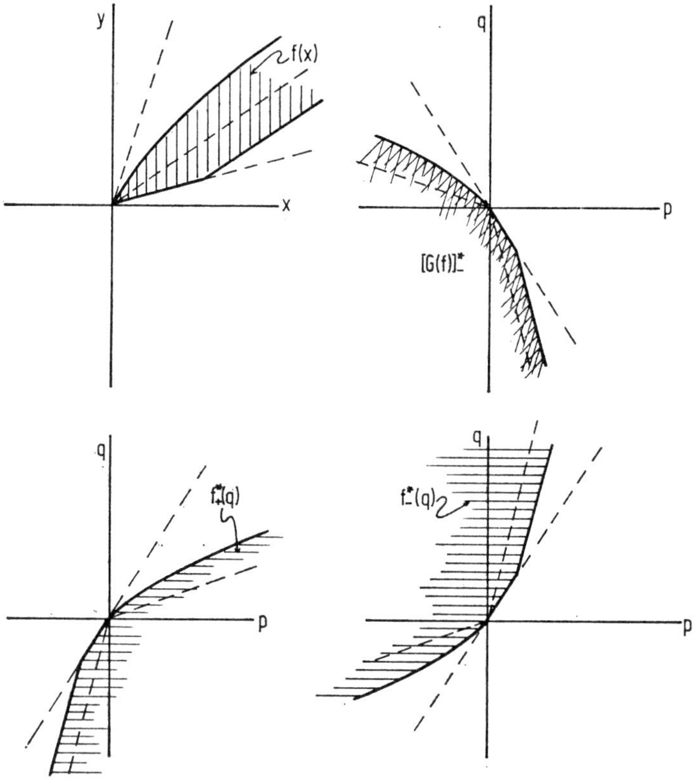

The behavior of f in the infinite is characterized by f_o, the cone opening of f, which by definition is not empty. If the behavior of a convex process near the origin and in the infinite is identical, then the cone opening and cone closure of the process coincide and the process reduces to a convex cone process, which behaves superlinearly.

It may be noticed that the behavior of a convex process near the origin determines the behavior at the infinite of the adjoint process, and vice versa. This important duality property follows from property 10.4.1, from which it can be deduced that the cone closure of a convex process determines the cone opening of the adjoint process, and the cone opening of a convex process determines the cone closure of the adjoint process.

Property 10.4.1. (*duality for cone closures and cone openings*)

Let $f: R^m \to R^n$ be a convex cone-interior process and let $f_+^*: R^{n*} \to R^{m*}$ be its upper adjoint process. Then:

1. $[G(f_o)]_-^* = [\text{Conint } G(f)]_-^* = \text{Cl Cone}[G(f)]_-^* = G[(-f_+^*)_c^{-1}]$;
   $[G(f_c)]_-^* = [\text{Cl Cone } G(f)]_-^* = \text{Conint } [G(f)]_-^* = G[(-f_+^*)_o^{-1}]$.
2. $R(f) = R(f_o) \quad \Leftrightarrow (f_+^*)^{-1}(0) = (f_+^*)_c^{-1}(0)$;
   $R(f_+^*) = R[(f_+^*)_o] \quad \Leftrightarrow \quad f^{-1}(0) = f_c^{-1}(0)$.

Proof

1. By definition $G(f_+^*) = \{(q, p) \mid (-p, q) \in [G(f)]_-^*\}$; therefore

$$G[(-f_+^*)^{-1}] = \{(p, q) \mid (p, q) \in [G(f)]_-^*\} = [G(f)]_-^*.$$

The first line of the statement follows from the definitions of f_c and f_o, and from property 9.3.1.2; the second line results when one applies the polar operation to property 9.3.1.2.

2. Assume that $R(f) \neq R(f_o)$; then a hyperplane $H(0, \bar{q})$ exists such that the half-space $M_-(0, \bar{q}) \supseteq G(f_o)$ and $M_-(0, \bar{q}) \not\supseteq G(f)$ (see fig. 10.4.2). From the definition of a lower polar set and from (1) above, it follows that $(0, \bar{q}) \in [G(f_o)]_-^* = G[(-f_+^*)_c^{-1}]$; also $(0, \bar{q}) \notin [G(f)]_-^* = G[(-f_+^*)^{-1}]$. Therefore $\bar{q} \in (f_+^*)_c^{-1}(0)$ and $\bar{q} \notin (f_+^*)^{-1}(0)$, from which it can be seen that $(f_+^*)^{-1}(0) \neq (f_+^*)_c^{-1}(0)$.

The converse statement follows by reversing the reasoning. The second equivalence follows from a similar reasoning, in which the second line under 1 is used. □

Fig. 10.4.2. $R(f) \neq R(f_o) \Leftrightarrow (f_+^*)^{-1}(0) \neq (f_+^*)_c^{-1}(0)$

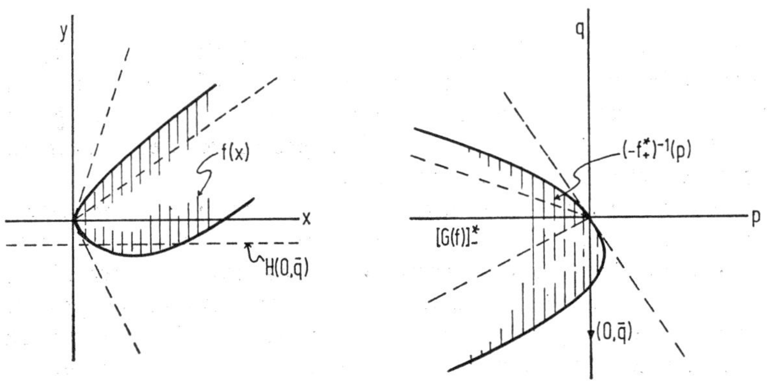

Property 10.4.2. (*convex cone-interior processes*)

Let $f: R^m \to R^n$ be a convex cone-interior process. Then:
1. f^{-1} is a convex cone-interior process;
2. f_+^* and f_-^* are convex cone-interior processes;
3. $f_o(0)$ is the recession cone of $f(x)$, for every x;
4. f is a convex cone process if and only if $f = f_c = f_o$;
5. f is a linear process if and only if $f = f_c = f_o$, $D(f) = R^m$ and $f(0) = \{0\}$.

Proof
1. f^{-1} is a convex process (see property 10.3.1.5); the other conditions are met by definition.

2. f_+^* and f_-^* are convex processes (see property 10.3.1.6). From their definitions it follows that $R(f_+^*)$ is a cone if and only if $R(f_-^*)$ is a cone, and that $f_+^*(0) = (f_+^*)_c(0)$ if and only if $f_-^*(0) = (f_-^*)_c(0)$. The same is true for the inverses.

Therefore, the statement follows from property 10.4.1.2, if it can be shown that $R(f)$ is a cone is equivalent to $R(f) = R(f_o)$:
Assume that $R(f)$ is not a cone; since $G(f_o)$ is a cone and $R(f_o) = \text{Proj}_2\, G(f_o)$ is a cone, $R(f) \neq R(f_o)$.

Conversely, suppose that $R(f) \neq R(f_o)$. Since $G(f_o) \subseteq G(f)$ and $G(f_o)$ is non-empty, $G(f)$ contains zero. According to property 8.3.1.1, Conint $G(f)$ equals the recession cone of $G(f)$, $0^+ G(f)$. As $R(f_o) = \text{Proj}_2\, G(f_o) = \text{Proj}_2 [0^+ G(f)]$, the existence of an element $\bar{y} \in R(f)$ such that $\bar{y} \notin R(f_o)$ implies that a $\lambda > 0$ exists such that $\lambda \bar{y} \notin R(f)$. Therefore $R(f)$ is not a cone. It follows that $R(f)$ is a cone if and only if $R(f) = R(f_o)$.

3. According to property 8.3.1.1, $G(f_o) = 0^+ G(f)$. The set of directions parallel to R^n in which $G(f)$ is unbounded is equal to $G(f_o) \cap \{(0, y) \mid y \in R^n\} = f_o(0)$, which is thus the recession cone of $\{y \mid (x, y) \in G(f)\} = f(x)$, for every $x \in D(f)$.

4. Let f be a convex cone process, then $G(f)$ is a cone and from property 8.3.1.1 it follows that $G(f) = \text{Conint}\, G(f) = \text{Cl Cone}\, G(f)$, implying that $f = f_o = f_c$. It can be easily seen that the conditions of a convex cone-interior process have been fulfilled (see property 10.3.2.4). Let f be a convex cone-interior process and $f = f_o = f_c$. Then $G(f)$ is a cone, implying that f is a convex cone process.

5. Let f be a linear process, then $G(f)$ is a subspace that meets the given conditions of a convex cone-interior process.

Let f be a convex cone-interior process; $f = f_o = f_c$ implies that f is a convex cone process. $D(f) = R^m$ and the superadditivity of f imply that $f(x)+f(-x) \subseteq f(x-x) = f(0)$, which is equal to $\{0\}$. It follows that for every x, $f(x)$ consists of a single element and $f(-x) = -f(x)$. Therefore f is closed under the operations of addition and scalar multiplication, i.e. f is a linear process. □

It may be noticed from (2) above that a convex process $f: R^m \to R^n$ is a convex cone-interior process if and only if $D(f)$, $R(f)$, $D(f_+^*)$ and $R(f_+^*)$ are all cones.

In order to apply algebraic operations on convex cone-interior processes, it is necessary that they have the same orientation. Two convex cone-interior processes f_1 and f_2 from R^m into R^n are said to have the *same orientation* if, for all $z_1 \in 0^+ G(f_1)$ and $z_2 \in 0^+ G(f_2)$, $z_1 + z_2 = 0$ implies that $z_1 = 0$ and $z_2 = 0$ and if $\text{Int } 0^+ [G(f_1) \cap G(f_2)] \neq \phi$. If two processes have the same orientation, their images cannot be unbounded in opposite directions (see fig. 10.4.3).

Fig. 10.4.3. Two processes, not having the same orientation

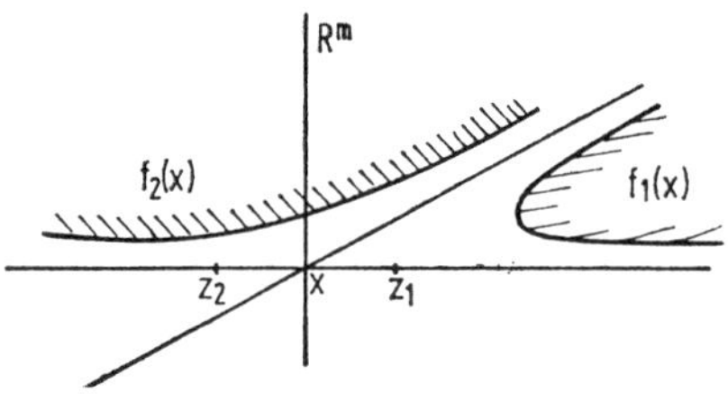

Property 10.4.3. (combinatorial operations)

Let f_1 and f_2 be convex cone-interior processes from R^m to R^n having the same orientation. Then the following processes are also convex cone-interior processes:

$f_1 + f_2$; $\quad f_1 \# f_2$;
$f_1 \wedge f_2$; $\quad f_1 \vee f_2$;
λf_1, for any scalar λ.

Proof

Addition: The set $G(f_1+f_2)$ may be constructed from $G(f_1)$ and $G(f_2)$ as follows. Define the sets K_1 and K_2 in $R^m \times R^m \times R^n$ by

$$K_1 := \{(x_1, 0, y_1) \mid (x_1, y_1) \in G(f_1)\};$$
$$K_2 := \{(0, x_2, y_2) \mid (x_2, y_2) \in G(f_2)\}.$$

Both sets are closed and convex. As f_1 and f_2 have the same orientation, $K_1 + K_2$ does not contain lines, or has zero lineality, and is therefore closed (see property 8.3.4). The sum of convex sets is also convex (see property 8.3.3).

Consider $K := \{K_1 + K_2\} \cap \{(x_1, x_2, z) \mid x_1 = x_2\}$: K is closed and convex. The image of K under the projection $(x_1, x_2, z) \to (x_1, z)$, $\text{Proj}_{1,3} K$, is closed: one can associate with every element $(\bar{x}_1, \bar{z}) \in \{(x_1, z) \mid \forall \varepsilon > 0,\ B[(x_1, z), \varepsilon] \cap \text{Proj}_{1,3} K \neq \phi\} = \text{Cl Proj}_{1,3} K$, an element $(\bar{x}_1, \bar{x}_2, \bar{z}) \in K$, because $\bar{x}_1 = \bar{x}_2$ and $\text{Proj}_2 K = D(f_2) = R(f_2^{-1})$ is a cone. Thus $\text{Cl Proj}_{1,3} K = \text{Proj}_{1,3} K$. The projection of a convex set is also convex. Therefore, $\text{Proj}_{1,3} K = \{(x_1, y_1 + y_2) \mid (x_1, y_1) \in G(f_1) \text{ and } (x_1, y_2) \in G(f_2)\} = G(f_1 + f_2)$ is a closed and convex set, implying that $(f_1 + f_2)$ is a convex process. Moreover $D(f_1 + f_2) = D(f_1) \cap D(f_2)$ is a cone, as is $R(f_1 + f_2) = \text{Proj}_3 K = R(f_1) + R(f_2)$, according to property 8.3.3.

Finally, as the partial addition of Cone $G(f_1)$ and Cone $G(f_2)$ is a convex cone, (from property 8.3.3), and as $(f_1 + f_2)$ is point-closed, (from property 10.3.1.3), and as by definition, $f_1(0) = (f_1)_c(0)$ and $f_2(0) = (f_2)_c(0)$, it follows that $(f_1 + f_2)(0) = ((f_1)_c + (f_2)_c)(0) = (f_1 + f_2)_c(0)$. This is also true for the inverse since, according to property 10.3.1.5, $(f_1 + f_2)^{-1}$ is also a convex process. It follows then that $(f_1 + f_2)$ is a convex cone-interior process.

Inverse addition: An analogous reasoning can be used to construct the set $G(f_1 \# f_2)$ from the sets $G(f_1)$ and $G(f_2)$. In this case the sets K_1 and K_2 are defined in $R^m \times R^n \times R^n$.

Conjunction: $G(f_1 \wedge f_2) = G(f_1) \cap G(f_2)$. It can be seen that $G(f_1 \wedge f_2)$ is a closed and convex set, hence $f_1 \wedge f_2$ is a convex process. Moreover, $D(f_1 \wedge f_2) = D(f_1) \cap D(f_2)$ and $R(f_1 \wedge f_2) = R(f_1) \cap R(f_2)$ are cones. Finally, as Cone $G(f_1) \cap$ Cone $G(f_2) = \text{Cone}[G(f_1) \cap G(f_2)]$ it follows from the definition of a cone closure of a convex process that $(f_1 \wedge f_2)(0) = ((f_1)_c \wedge (f_2)_c)(0) = (f_1 \wedge f_2)_c(0)$. This is valid again for the inverse; hence it follows that $(f_1 \wedge f_2)$ is a convex cone-interior process.

Disjunction: $G(f_1 \vee f_2) = G(f_1) + G(f_2)$. As f_1 and f_2 have the same orientation, $G(f_1 \vee f_2)$ is a closed and convex set; therefore $f_1 \vee f_2$ is a convex process. The other conditions can be analogously derived from the above reasoning.

Scalar multiplication: For a given λ, the set $G(\lambda f_1) = \{(x, \lambda y) | (x, y) \in G(f_1)\}$ is closed and convex because $G(f_1)$ is closed and convex. Therefore λf_1 is a convex process. The other conditions are expressed in cones, which are invariant under scalar multiplication. □

Property 10.4.4. (*composition*)

Let $f_1 : R^l \to R^m$ and $f_2 : R^m \to R^n$ be convex cone-interior processes; then the product $f_2 f_1$ is also a convex cone-interior process.

Proof
The set $G(f_2 f_1)$ may be constructed from $G(f_1)$ and $G(f_2)$ as follows: define the sets K_1 and K_2 in $R^l \times R^m \times R^m \times R^n$ by

$$K_1 := \{(x_1, y_1, 0, 0) | y_1 \in f_1(x_1)\};$$
$$K_2 := \{(0, 0, y_2, z_2) | z_2 \in f_2(y_2)\}.$$

The sum $K_1 + K_2$ is closed since for all $k_1 \in 0^+ K_1$ and $k_2 \in 0^+ K_2$, $k_1 + k_2 = 0$ implies $k_1 = 0$ and $k_2 = 0$; closure follows from property 8.3.4. The sum is also convex.

Consider $K := (K_1 + K_2) \cap \{(x, y_1, y_2, z) | y_1 = y_2\}$; K is closed and convex. The image of K under the projection $(x, y_1, y_2, z) \to (x, z)$, $\text{Proj}_{1,4} K$, is closed and convex. Closure follows from the fact that one can associate with every limit point $(\bar{x}, \bar{z})$ of a sequence $(x_t, z_t) \in \text{Proj}_{1,4} K$, an element $(\bar{x}, \bar{y}, \bar{y}, \bar{z}) \in K$, because a certain y_t exists such that $y_t \in f_1(x_t)$ and $z_t \in f_2(y_t)$ and both $R(f_1)$ and $D(f_2)$ are cones. As $\text{Proj}_{1,4} = \{(x, z) | \exists y \in f_1(x) : z \in f_2(y)\} = G(f_2 f_1)$, the multifunction $f_2 f_1$ is a convex process. Moreover, as $R(f_2)$ is a cone, $R(f_2 f_1)$ is a cone; as $D(f_1)$ is a cone, $D(f_2 f_1)$ is a cone.

Finally, as $f_2(0) = (f_2)_c(0)$ and both $f_1(0)$ and $f_2 f_1(0)$ are cones, it follows that $f_2 f_1(0) = [(f_2)_c (f_1)_c](0) = (f_2 f_1)_c(0)$. The same holds for the inverse of $f_2 f_1$. It follows that $(f_2 f_1)$ is a convex cone-interior process. □

Property 10.4.5. (*inverse operations*)

Let f_1 and f_2 be convex cone-interior processes from R^m into R^n having the same orientation. Then the inverse operation has the following properties:

1. $(f_1 + f_2)^{-1} = f_1^{-1} \# f_2^{-1}$;
2. $(f_1 \# f_2)^{-1} = f_1^{-1} + f_2^{-1}$;
3. $(f_1 \wedge f_2)^{-1} = f_1^{-1} \wedge f_2^{-1}$;

4. $(f_1 \vee f_2)^{-1} = f_1^{-1} \vee f_2^{-1}$;
5. $(\lambda f)^{-1}(y) = f^{-1}(y/\lambda)$, for some $\lambda > 0$.

Proof
1. $G[(f_1+f_2)^{-1}] = \{(y, x) \mid y \in f_1(x) + f_2(x)\}$
$= \{(y, x) \mid \exists (x, y_1) \in G(f_1), \exists (x, y_2) \in G(f_2) : y = y_1 + y_2\}$
$= \{(y, x) \mid \exists y_1, y_2 : x \in f_1^{-1}(y_1) \cap f_2^{-1}(y_2) \text{ and } y = y_1 + y_2\}$
$= G(f_1^{-1} \# f_2^{-1})$.

2. The reverse argument can be applied.

3. $G[(f_1 \wedge f_2)^{-1}] = \{(y, x) \mid (x, y) \in G(f_1) \cap G(f_2)\}$
$= G(f_1^{-1}) \cap G(f_2^{-1}) = G(f_1^{-1} \wedge f_2^{-1})$.

4. $G[(f_1 \vee f_2)^{-1}] = \{(y, x) \mid (x, y) \in G(f_1) + G(f_2)\}$
$= G(f_1^{-1}) + G(f_2^{-1}) = G(f_1^{-1} \vee f_2^{-1})$.

5. $\{x \mid x \in (\lambda f)^{-1}(y)\} = \{x \mid (x, y) \in G(\lambda f)\} = \{x \mid (x, y/\lambda) \in G(f)\}$
$= \{x \mid x \in f^{-1}(y/\lambda)\}$. □

Property 10.4.5.2 indicates why the operation # is called 'inverse addition'. From properties 10.4.2 and 10.4.3 it follows that the above multifunctions are also convex cone-interior processes.

Property 10.4.6. (adjoint operations)

Let f_1 and f_2 be convex *cone-interior* processes from R^m into R^n having the same orientation. Let * indicate either the upper adjoint or the lower adjoint operation. Then:

1. $(f_1 + f_2)^* \supseteq f_1^* \mathring{+} f_2^*$;
2. $(f_1 \# f_2)^* \supseteq f_1^* \mathring{\#} f_2^*$;
3. $(f_1 \wedge f_2)^* = f_1^* \mathring{\vee} f_2^*$;
4. $(f_1 \vee f_2)^* = f_1^* \mathring{\wedge} f_2^*$;
5. $(\lambda f)^*(q) = f^*(\lambda q)$, for $\lambda > 0$.

It can be proven that the inclusions expressed in 1. and 2. can be replaced by equalities, but this will not be attempted here.

Proof
The property is demonstrated for the upper adjoint operator; a similar argument can be applied to prove the proposition for the lower adjoint operator, except that some signs should then be reversed.

1. $G[(f_1 + f_2)^*_+] = \{(q, p) | (-p, q) \in [G(f_1 + f_2)]^*_-\}$

$= \{(q, p) | \forall (x_1, y_1) \in G(f_1), \forall (x_2, y_2) \in G(f_2): x = x_1 = x_2, y = y_1 + y_2$ and $-px + qy \leqq 1\}$

$\supseteq \bigcup_{\lambda \in [0,1]} (q, p) | \exists q_1, q_2, p_1, p_2: \; q = (1-\lambda) q_1 = \lambda q_2, \; p = (1-\lambda) p_1 + \lambda p_2,$
$\forall (x_1, y_1) \in G(f_1), \forall (x_2, y_2) \in G(f_2): (1-\lambda) q_1 y_1 - (1-\lambda) p_1 x_1 \leqq (1-\lambda)$ and $\lambda q_2 y_2 - \lambda p_2 x_2 \leqq \lambda\}$

$= \bigcup \{(q, p) | \exists q_1, q_2, p_1, p_2: q = (1-\lambda) q_1 = \lambda q_2, p = (1-\lambda) p_1 + \lambda p_2,$
$\forall (x_1, y_1) \in G(f_1): -p_1 x_1 + q_1 y_1 \leqq 1$ and
$\forall (x_2, y_2) \in G(f_2): -p_2 x_2 + q_2 y_2 \leqq 1\}$

$= \bigcup \{(q, p) | \exists q_1, q_2, p_1, p_2: q = (1-\lambda) q_1 = \lambda q_2, p = (1-\lambda) p_1 + \lambda p_2,$
$p_1 \in f^*_{1+}(q_1), p_2 \in f^*_{2+}(q_2)\}$

$= \bigcup \{q, p) | \exists q_1, q_2: p = (1-\lambda) f^*_{1+}(q_1) + \lambda f^*_{2+}(q_2)$ and
$q = (1-\lambda) q_1 = \lambda q_2\} = G(f^*_{1+} \mathring{+} f^*_{2+})$.

2. $G[(f_1 \# f_2)^*_+] = \{(q, p) | (-p, q) \in [G(f_1 \# f_2)]^*_-\}$

$= \{(q, p) | \forall (x_1, y_1) \in G(f_1), \forall (x_2, y_2) \in G(f_2):$
$x = x_1 + x_2, y = y_1 = y_2$ and $-px + qy \leqq 1\}$

$\supseteq \bigcup_{\lambda} \{(q, p) | \exists q_1, q_2, p_1, p_2: q = (1-\lambda) q_1 + \lambda q_2, p = (1-\lambda) p_1 = \lambda p_2,$
$\forall (x_1, y_1) \in G(f_1), \forall (x_2, y_2) \in G(f_2):$
$(1-\lambda) q_1 y_1 - (1-\lambda) p_1 x_1 \leqq (1-\lambda)$ and $\lambda q_2 y_2 - \lambda p_2 x_2 \leqq \lambda$

$= \bigcup \{(q, p) | \exists q_1, q_2, p_1, p_2: q = (1-\lambda) q_1 + \lambda q_2, p = (1-\lambda) p_1 = \lambda p_2,$
$p_1 \in f^*_{1+}(q_1)$ and $p_2 \in f^*_{2+}(q_2)\}$

$= G(f^*_{1+} \mathring{\#} f^*_{2+})$.

3. $G[(f_1 \wedge f_2)^*_+] = \{(q, p) | (-p, q) \in [G(f_1 \wedge f_2)]^*_-\}$

$= \{(q, p) | (-p, q) \in [G(f_1) \cap G(f_2)]^*_-\}$

$= \{(q, p) | (-p, q) \in [G(f_1)]^*_- \mathring{+} [G(f_2)]^*_-\}$, from property 9.3.4.2,

$= G(f^*_{1+}) \mathring{+} G(f^*_{2+}) = G(f^*_{1+} \mathring{\vee} f^*_{2+})$. (See section 10.1)

4. $G[(f_1 \vee f_2)^*_+] = \{(q, p) | (-p, q) \in [G(f_1 \vee f_2)]^*_-\}$

$= \{(q, p) | (-p, q) \in [G(f_1) + G(f_2)]^*_-\}$

$= \{(q, p) | (-p, q) \in [G(f_1)]^*_- \mathring{\cap} [G(f_2)]^*_-\}$, from property 9.3.4.1,

$= G(f^*_{1+}) \mathring{\cap} G(f^*_{2+}) = G(f^*_{1+} \mathring{\wedge} f^*_{2+})$. (See section (10.1.)

5. $G[(\lambda f)^*_+] = \{(q, p) | (-p, q) \in [G(\lambda f)]^*_-\}$

$= \{(q, p) | \forall (x, y) \in G(f): -px + q(\lambda y) \leqq 1\}$

$= \{(q, p) | (-p, \lambda q) \in [G(f)]^*_-\}$. □

From properties 10.4.1, 10.4.2 and 10.4.3 it follows that the class of convex cone-interior processes is also closed under the convex algebraic operations defined in section 10.1.

Property 10.4.7. (*adjoint operations*)

Let f_1 and f_2 be convex *cone* processes from R^n into R^m having the same orientation. Let $*$ indicate either the upper or the lower adjoint operation. Then:

1. $(f_1 + f_2)^* \supseteq f_1^* + f_2^*$;
2. $(f_1 \# f_2)^* \supseteq f_1^* \# f_2^*$;
3. $(f_1 \wedge f_2)^* = f_1^* \vee f_2^*$;
4. $(f_1 \vee f_2)^* = f_1^* \wedge f_2^*$;
5. $(\lambda f)^* = \lambda f^*$.

Proof

This property follows from the fact that the graphs are cones and that the operations in property 10.4.6 can be reduced in accordance with property 10.1.2. Compare this result with that for sets in section 9.3 (properties 9.3.4 and 9.3.5), and with the result for Rockafellar's (1972) definition of an adjoint in property 10.3.3. □

10.5. CONVEX STAR PROCESSES AND CONVEX AUREOLE PROCESSES

For most applications a convex cone-interior process is a too general model and must therefore be confined to a starred or aureoled process, as defined in section 10.2. If a convex cone-interior process is starred, it is said to be a *convex star process*; if it is aureoled, it is called a *convex aureole process* (see fig. 10.5.1).

Fig. 10.5.1. A convex aureole process f and its convex star process f^*_-

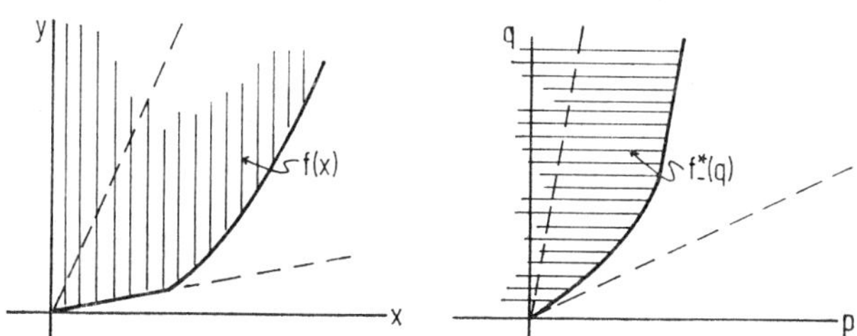

In economic theory, many production models or consumption models are based on convex star or aureole processes (see chapters 3 and 4 of this book). Some properties of these processes are given below.

Property 10.5.1. (convex star processes)

Let $f: R^m \to R^n$ be a convex star process. Then:
1. $f(x) = f(\text{Star } x) = \text{Star } f(x)$;
2. $f_o(0) \subseteq f(x) \subseteq f(\lambda x) \subseteq \lambda f(x)$, for $\lambda \geqq 1$;
3. f^{-1} is a convex aureole process;
4. f_+^* is a convex aureole process;
5. ${}_-^*f$ is a point-starred and graph-closed multifunction.

Proof
1. $f(x) \supseteq f(\text{Star } x) = \bigcup \{f(\lambda x) | 0 \leqq \lambda \leqq 1\}$:
Suppose that some $\bar{y} \in f(\bar{\lambda} x)$, with $0 < \bar{\lambda} < 1$, is not an element of $f(x)$. Then $\bar{\lambda} x \in f^{-1}(\bar{y})$ and $x \notin f^{-1}(\bar{y})$. This contradicts f^{-1} being point-aureoled (see section 10.2), by which $x = \bar{\lambda}^{-1}(\bar{\lambda} x) \in f^{-1}(\bar{y})$, as $\bar{\lambda}^{-1} > 1$.

It is evident that $f(x) \subseteq f(\text{Star } x)$, hence $f(x) = f(\text{Star } x)$.
$f(x) = \text{Star}[f(x)]$ follows from f being point-starred.

2. $G(f_o)$ is the recession cone of $G(f)$, from the definition of the cone opening f_o (see section 10.3) and property 8.3.1.1. As $G(f_o) \subseteq G(f)$ and $0 \in f(0)$, it follows that $f_o(0) \subseteq f(0) \subseteq f(x)$.

Further, $f(x) \subseteq f(\lambda x)$, for $\lambda \geqq 1$, follows from f^{-1} being point-aureoled.

Finally, $f(\lambda x) \subseteq \lambda f(x)$, for $\lambda \geqq 1$. Choose any $y \in f(\lambda x)$; since $(\lambda x, y) \in G(f)$ and $(0, 0) \in G(f)$, it follows from the convexity of $G(f)$ that $(x, y/\lambda) \in G(f)$, or $y \in \lambda f(x)$.

3. By definition.

4. Let $p \in f_+^*(q)$; then $\forall x, \forall y \in f(x) : px \geqq qy - 1$. Since $f^{-1}(y)$ is point-aureoled, it follows that for all $\lambda \geqq 1$: $p(\lambda x) = (\lambda p)x \geqq qy - 1$. Therefore $(\lambda p) \in f_+^*(q)$ and f_+^* is point-aureoled. An analogous argument shows that $(f_+^*)^{-1}$ is point-starred; therefore f_+^* is aureoled.

The lower adjoint process f_-^* is also aureoled, but is not relevant in the case of a convex star process (see fig. 10.5.2).

5. ${}_-^*f$ being point-starred follows from the definition (section 10.1) and property 9.2.3.4 (see also fig. 10.2.3). That ${}_-^*f$ is graph-closed follows from properties 10.1.3.2 and 10.3.1.4. □

This result for convex star processes can also be obtained for convex aureole processes (property 10.5.2). The proofs are similar to those given above.

Property 10.5.2. (*convex aureole processes*)

Let $f: R^m \to R^n$ be a convex aureole process. Then:

1. $f(x) = f(\text{Aur } x) = \text{Aur } f(x)$;
2. $f_o(0) \supseteq f(x) \supseteq f(\lambda x) \supseteq \lambda f(x)$, for $\lambda \geqq 1$;
3. f^{-1} is a convex star process;
4. f^*_- is a convex star process;
5. ${}^*_+f$ is a point-aureoled and graph-closed multifunction.

The class of convex star and convex aureoled processes is closed under all operations for which the class of convex cone-interior processes is closed (see section 10.4).

It should also be noticed that non-negative monotone processes are either starred or aureoled, according to property 10.2.6. These kinds of processes are frequently used.

Finally, for a convex star or a convex aureole process a relation can be established between its polar multifunction and its adjoint process. This is shown in the following property.

Property 10.5.3. (*adjoint and polar multifunctions*)

Let $f: R^m \to R^n$ be a convex star process. Then:

$$f^*_+(q) = \bigcap_y \{(qy-1)\ {}^*_+(f^{-1})\,(y) \mid qy > 1\}, \text{for any } q \in R^{n*};$$

$${}^*_-f(x) = \text{Conv} \bigcup_p \{(px+1)^{-1}\ (f^*_+)^{-1}(p) \mid px > -1\}, \text{ for some } x \in R^m.$$

Let $f: R^m \to R^n$ be a convex aureole process. Then:

$$f^*_-(q) = \bigcap_y \{(qy+1)\ {}^*_-(f^{-1})\,(y) \mid qy > -1\}, \text{ for some } q \in R^{n*};$$

$${}^*_+f(x) = \text{Conv} \bigcup_p \{(px-1)^{-1}\ (f^*_-)^{-1}\,(p) \mid px > 1\}, \text{ for some } x \in R^m.$$

Proof (see fig. 10.5.2)

$$\begin{aligned} f^*_+(q) &= \{p \mid \forall x, \forall y \in f(x) : px \geqq qy-1\} \\ &= \{p \mid \forall y, \forall x \in f^{-1}(y) : px \geqq qy-1\} \\ &= \bigcap_y \{p \mid \forall x \in f^{-1}(y) : px \geqq qy-1\}. \end{aligned}$$

Given some q, the following three cases can be distinguished:

a. $qy-1=0$, implying that for such a y:
$\{p \mid \forall x \in f^{-1}(y) : px \geqq qy-1\} = \{p \mid \forall x \in f^{-1}(y) : px \geqq 0\} = [f^{-1}(y)]^0_+$.

b. $qy-1<0$, implying that (see properties 9.2.3.3 and 9.2.5.3):
$\{p \mid \forall x \in f^{-1}(y) : (-p)x \leqq (1-qy)\} =$

$= \{p \mid \forall x \in \mathrm{Star} f^{-1}(y) : (-p)x \leqq (1-qy)\}$
$= \{p \mid \forall x \in \mathrm{Cone} f^{-1}(y) : (-p)x \leqq 0\}$
$= -[f^{-1}(y)]^0_- = [f^{-1}(y)]^0_+.$

c. $qy-1>0$; then $\{p \mid \forall x \in f^{-1}(y) : px \geqq (qy-1)\}$
$= (qy-1)\,\{p \mid \forall x \in f^{-1}(y) : px \geqq 1\} = [f^{-1}(y)]^*_{(qy-1)+}.$

According to property 9.2.1,

$$\lim_{qy \to 1} [f^{-1}(y)]^*_{(qy-1)+} = [f^{-1}(y)]^0_+.$$

Because f is starred, f^{-1} is aureoled and, from property 9.2.5.2, $[f^{-1}(y)]^*_+ \subseteq [f^{-1}(y)]^0_+$. Together this implies that:

$$f^*_+(q) = \bigcap_y (qy-1)\{p \mid \forall x \in f^{-1}(y) \colon px \geqq 1\}, \quad \text{for } qy > 1$$
$$= \bigcap_y \{(qy-1)\,{}^*_+(f^{-1})(y) \mid qy > 1\}.$$

If f is starred, then f^*_+ is aureoled (see property 10.5.1.4). The upper polar multifunction is equal to:

$${}^*_+(f^*_+)(q) = [\bigcap_y (qy-1)\,{}^*_+(f^{-1})(y)]^*_+, \qquad \text{for } qy > 1.$$

From properties 9.3.4.5 and 9.2.2.5:

$${}^*_+(f^*_+)(q) = \mathrm{Conv} \bigcup_y [(qy-1)^{-1}\,{}^*_+(f^{-1})(y)]^*_+ =$$
$$= \mathrm{Conv} \bigcup_y [(qy-1)^{-1} f^{-1}(y)], \qquad \text{for } qy > 1.$$

Substitute $\bar{f}(x)$ for $f^*_+(q)$, both being aureoled, to get

$${}^*_+\bar{f}(x) = \mathrm{Conv} \bigcup_p [(px-1)^{-1}(\bar{f}^*_-)^{-1}(p)], \qquad \text{for } px > 1.$$

Similar arguments can be put forward when f is aureoled. □

If f is a non-negative convex cone process, then the property outlined above can be simplified. It has been shown (Ruys, 1972) that for the convex polyhedral process $f\colon R^n_+ \to R^n_+$, defined by $f(x) := \{y \mid Ay \leqq x\}$, in which matrix A is regular and has a dominant diagonal, the following relation between the adjoint and the polar multifunction is valid:

$$f^*_+(q) \quad = \bigcap_y \{{}^*_+(f^{-1})(y) \cap R^{n*}_+ \mid qy = 1 \quad \text{and} \quad y \geqq 0\};$$
$$(f^{-1})^*_-(p) = \bigcap_x \{{}^*_-f(x) \cap R^{n*}_+ \mid px = 1 \quad \text{and} \quad x \geqq 0\}.$$

In this problem, the adjoint $f^*_+(\bar{q}) = \{p \in R^{n*}_+ \mid A^* p \geqq \bar{q}\}$ determines the set of constraints in a linear programming problem which is the dual

Fig. 10.5.2. A convex star process, f, and its adjoint process, f^*_+; $^*_+f^{-1}$ is the upper polar process of the convex aureole process f^{-1}.

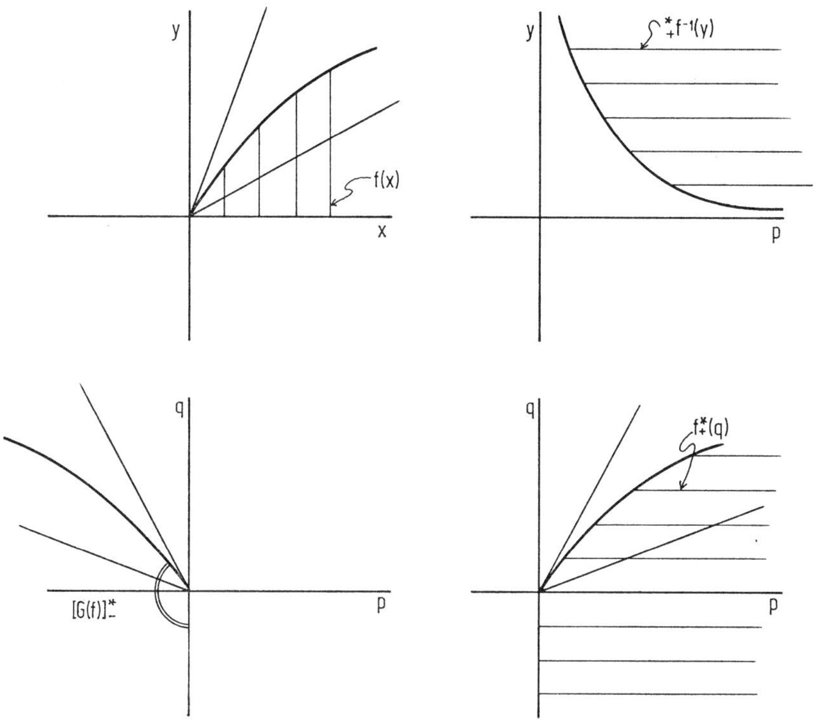

problem of: $\max\{\bar{q}y \mid y \in f(\bar{x})\}$. This property of adjoints has been pointed out above in a more general context (see the duality theorem 10.3.6).

A diagram relating the various adjoint, polar and inverse multifunctions can be found in section 10.1.

Summary

The classical model of General Equilibrium, as given by Walras, treats an economy in which a number of consumers and producers are present, each having their own preferences and production-capacities. In this economy, decentralization of decisions is made possible by allotting to each participant his own choice-set (determined by his income, for example) from which he chooses a subjective best element. It has been shown that, under certain conditions, an allocation can then be realized which holds an equilibrium between possibilities and wishes for all participants, and which satisfies some optimality conditions. The classical model, however, is meant only for an exchange economy with private goods, i.e. commodities which are used exclusively by one participant and which involve no external effects for the other participants. Although it is possible to also introduce public goods (used by all participants simultaneously) into the model, serious problems arise in connection with the formation of decisions.

In this study an attempt is made to solve this problem and to propose an organization of individual decisions in a distribution economy which satisfies the equilibrium conditions mentioned above. The proposal is made here to separate decisions on the level of private goods from decisions on the level of public goods, by introducing two standards of value. Each participant is allotted an income for private goods and a choice-set of valuation-prices for public goods, from which he can choose an element indicating priorities between several public goods. An equilibrium is said to be obtained if the so-determined valuations, summed over the individuals, are proportional to the costs of providing that bundle of public goods, and if the proportion between the share of income bestowed on private goods and the share of income bestowed on public goods causes no dis-equilibrium in the production sector. Such an equilibrium is called a two-level price equilibrium. The organization can be extended to several levels, so that local public goods can also be considered.

The method used is made possible by applying a duality transformation on the economy, which expresses all relations characterizing a participant in terms of prices, rather than in terms of quantities. This duality operation (which is perfectly satisfactory only in a convex environment) functions as a translator expressing information in quantities in terms of information in prices. Given the character of public goods, the individual participant must make a choice in terms of valuation-prices, and collectively a decision is made about the quantities. The mirror image is present in the case of the allocation of private goods.

Extensive attention is therefore devoted in this study to the properties and effects of duality operations. In the second part of the book, these transformations are treated in connection with sets, and a generalization is given of these operations applied to multifunctions. The effects of these on algebraic operations, such as addition, are investigated, and algebraic operations for sets and multifunctions in the dual space are defined, permitting aggregation also in the price space.

The economy in the price space derived via such a duality transformation is not essentially different from the economy in the quantity space. Each is a representation of the other. If, however, both systems exist without being related by a completely deterministic duality operation and a dynamic interpretation is given to the transformation, then an endogenous process of development can be explained for the whole of systems and relations which constitute an economy.

References

Aoki, M., 'Two Planning Algorithms for an Economy whith Public Goods', Discussion Paper No. 029, Kyoto Institute of Economic Research, 1970.

Arrow, K.J., *Social Choice and Individual Values*, John Wiley and Sons, New York, 1951.

Arrow, K.J., *The organization of economic activity: issues pertinent to the choice of market versus non-market allocation in the analysis and evaluation of public expenditures: the PPB system*, Joint Economic Committee of the Congress of the United States, 1969, p. 47–64.

Arrow, K.J., 'Political and Economic Evaluation of Social Effects and Externalities', in: *The Analysis of Public Output* (ed. J. Margolis), National Bureau of Economic Research, Columbia University Press, New York, 1970.

Arrow, K.J., and G. Debreu, 'Existence of Equilibrium for a Competitive Economy', *Econometrica, 22*, 1954, p. 265–290.

Arrow, K.J., and F.H. Hahn, *General Competitive Analysis*, Holden-Day, San Francisco, 1971.

Arrow, K.J., and L. Hurwicz, 'Gradient Method for Concave Programming', in: K.J. Arrow, L. Hurwicz and H. Uzawa, eds., *Studies in Linear and Non-Linear Programming*, Stanford, 1958 chs. 6, 8.

Arrow, K.J., and L. Hurwicz, 'Decentralization and Computation in Resource Allocation' in *Essays in Economics and Econometrics* (ed. R.W. Pfouts) University of North Carolina Press, Chapel Hill, N.C., 1960, p. 34–104.

Aumann, R.J., 'Markets with a continuum of traders', *Econometrica, 32*, 1964, p. 39–50.

Barone, E., 'Il ministerio della produzione nello stato colletivista', *Giornale degli Economisti*, 1908. Translated as 'The ministry of production in the collectivist state', in: *Collectivist Economic Planning* (ed. F.A. von Hayek), Routledge, London, 1935, p. 245–90.

Barone, E., 'On Public needs', translated in: *Classics in the Theory of Public Finance* (eds. R.A. Musgrave and A.T. Peacock), Macmillan, London, 1912.

Baumol, W.J., 'On taxation and the control of externalities', *American Economic Review, 62*, 1972, p. 307–322.

Baumol, W.J., and T. Fabian, 'Decomposition, Pricing for Decentralization and external economics', *Management Science, 11*, 1964, p. 1–32.

Berge, C., *Espaces topologiques*, Dunod, Paris, 1959; translated as: *Topological spaces*, Oliver & Boyd, Edinburgh, 1963.

Bergstrom, T.C., 'Collective choice and the Lindahl allocation method', mimeographed, Washington University, St. Louis, Mo, 1971.

Bergstrom, T.C., and R.P. Goodman, 'Private demands for public goods' *The American Economic Review, 63*, 1973, p. 280–296.
Bohm, P., 'On the theory of second best', *Review of Economic Studies, 34*, 1967, p. 301–314.
Bowen, H.R., 'The interpretation of voting in the allocation of economic resources', *Quarterly Journal of Economics, 58*, 1943, p. 27–48.
Buchanan, J.M., *The Demand and Supply of Public Goods*, Rand McNally, Chicago, 1968.

Camacho, A., 'Centralization and Decentralization of Decision Making Mechanisms: A General Model', *Jahrbuch der Wirtschaft Osteuropas, 3*, 1972, p. 45–66.
Chipman, J.S., L. Hurwicz, M.K. Richter and H.F. Sonnenschein, eds., *Preferences, Utility and Demand*, Harcourt Brace, New York, 1971.
Cobbenhagen, M.J.H., 'De economische orde', *Maandschrift Economie, 8*, 1943.
Cournot, A.A., *Recherches sur les principes mathématiques de la théorie des richesses*, 1838. Nouvelle édition (publiée par G. Lutfalla), Rivière, Paris, 1938.

Dalmulder, J.J.J., Plaats en betekenis van de wiskundige economie en de econometrie binnen het geheel der economische wetenschappen, Intree rede, Bergmans, Tilburg, 1950.
Dalmulder, J.J.J., 'Het wiskundig en het economisch denken', *Maandschrift Economie, 25*, 1960, p. 57–74.
Dantzig, G.B., 'Programming of Interdependent Activities: Mathematical Model' *Econometrica, 17*, 1949, p. 200–11; reprinted in: T.C. Koopmans (ed.), *Activity Analysis of Production and Allocation*, John Wiley, New York, 1951.
Dantzig, G.B., and P. Wolfe, 'Decomposition Principle for Linear Programs', *Operations Research, 8*, 1960, p. 101–111.
Dantzig, G.B., and P. Wolfe, 'The decomposition algorithm for linear programs', *Econometrica, 29*, 1961, p. 767–778.
Davis, O.A., and A.B. Whinston, 'Welfare economics and the theory of second best', *Review of Economic Studies, 32*, 1965, p. 1–14.
Debreu, G., *Theory of Value*, John Wiley and Sons, New York, 1959.
Debreu, G., 'New Concepts and techniques for equilibrium analysis', *International Economic Review, 3*, 1962, p. 257-273.
Debreu, G., and H. Scarf, 'A limit theorem on the core of an economy', *International Economic Review, 4*, 1963, p. 235–246.
Diamond, P.A., and J.A. Mirrlees, 'Optimal Taxation and Public Production', *The American Economic Review, 61*, 1971, pp. 8–27 and 261–278.
Dorfman, R., 'General Equilibrium with Public Goods', in: *Public Economics* (eds. J. Margolis and H. Guitton). Macmillan, London, 1969, p. 247–275.
Drees, W., en F. Th. Gubbi, *Overheidsuitgaven in theorie en praktijk*, Wolters Noordhoff, Groningen, 1968.
Drèze, J.H., and D. de la Vallée Poussin, 'A Tâtonnement Process for Public Goods', *The Review of Economic Studies, 38*, 1971, p. 133–150.
Drèze, J.H., 'Investment under private ownership: optimality, equilibrium and stability' in: *Allocation under Uncertainty: Equilibrium and Optimality*, Proceedings of a Conference held by the I.E.A. (ed. J.H. Drèze), Macmillan, London, 1974.
Downs, Anthony, *An Economic Theory of Democracy*, Harper and Row, New York, 1957.
Dupuit, Jules, 'De la mesure de l'utilité des travaux publics', *Annales des Ponts et*

Chaussées, *8*, 1844; translated 'On the measurement of the utility of public works' in: *Readings in Welfare Economics* (eds. K.J. Arrow and T. Scitovisky) A.E.A. Series, George Allen and Unwin, London, 1969, p. 255–283.

E.E.C., *De ontwikkeling van de overheidsfinanciën in de Lid-Staten van de Europese Gemeenschappen van 1957 tot 1966*, Verzameling Studies Serie Economische en Financiële Zaken nr. 8, Commissie van de Europese Gemeenschappen, Brussel, 1970.

Edgeworth, F.Y., *Mathematical Psychics*, Paul Kegan, London, 1881.

Ellman, M., *Soviet Planning Today: Proposals for an Optimally Functioning Economic System*, University Press, Cambridge, 1971.

Evers, J.J.M., *Linear Programming over an Infinite Horizon*, Tilburg University Press, 1973.

Fabre, F., 'Biens collectifs et biens à qualité variable', Paper presented at the European Meeting of the Econometric Society, Brussels, 1969.

Fan, K., 'Fixed point and minimax theorems in locally convex topological linear spaces', *Proceedings of the National Academy of Sciences, U.S.A.*, *38*, 1952, p. 121–126.

Fleming, W.H., *Functions of several variables*, Addison-Wesley Publishing Company, Reading, Mass., 1965.

Foley, D.K., 'Resource Allocation and the Public Sector', *Yale Economic Essays*, *7*, 1967, p. 45–98.

Foley, D., 'Lindahl's Solution and the Core of an Economy with Public Goods', *Econometrica*, *38*, 1970, p. 66–72.

Fourgeaud, Cl., 'Contribution à l'étude du rôle des administrations dans la théorie mathématique de l'équilibre et de l'optimum', *Econometrica*, *37*, 1969, p. 307–323.

Guesnerie, R., 'Le rôle de la fiscalité et du calcul économique public dans un modèle simple de second rang', Paper presented at the European Meeting of the Econometric Society, Oslo; Cepremap, Paris, 1973.

Hahn, F.H., 'On Optimum Taxation', *Journal of Economic Theory*, *6*, 1972, p. 96–106.

Hayek, F.A. von, 'The Present State of the Debate' in: *Collectivist Economic Planning* (ed. F.A. von Hayek), Routledge, London, 1935, p. 201–243.

Hayek, F.A., 'The Use of the Knowledge in Society', *The American Economic Review*, *35*, 1945, p. 519–530.

Head, J.G., 'Public Goods and Public Policy', *Public Finance*, *17*, 1962, p. 197–221.

Heal, G.M., 'Planning without Prices, *Review of Economic Studies*, *36*, 1969, p. 347–362.

Heal, G.M., 'Planning, Prices, and Increasing Returns' *Review of Economic Studies*, *38*, 1971, p. 281–294.

Henry, C., 'Differential equations with discontinuous right-hand side for planning procedures', *Journal of Economic Theory*, *4*, 1972, p. 545–551.

Hildenbrand, W., 'Existence of equilibria for economics with production and a measure space of consumers', *Econometrica*, *38*, 1970, p. 608–623.

Hotelling, H., 'Edgeworth's taxation paradox and the nature of demand and supply functions', *Journal of Political Economy*, *40*, 1932, p. 577-616.

Hurwicz, L., 'Optimality and Informational Efficiency in Resource Allocation Processes' in: *Mathematical Methods in the Social Sciences, 1959* (eds. K.J. Arrow, S. Karlin, P. Suppes), Stanford University Press Stanford, California, 1960, p. 27–46.

Hurwicz, L., 'Centralization and decentralization in economic processes', *Jahrbuch der Wirtschaft Osteuropas*, *3*, 1972, p. 87–113.

Hurwicz, L., 'The Design of Mechanisms for Resource Allocation', Richard T. Ely Lecture at the 85-th A.E.A. Meeting, *The American Economic Review*, *63*, 1973, p. 1–30.

Jacobson, S.E., 'Production correspondences', *Econometrica*, *38*, 1970, p. 754–771.

Johansen, L., 'Some Notes on the Lindahl Theory of Determination of Public Expenditures', *International Economic Review*, *4*, 1963, p. 346–358.

Johansen, L., *Public Economics*, North-Holland Publishing Company, Amsterdam, 1965.

Kaizuka, K., 'Public Goods and Decentralization of Production', *The Review of Economics and Statistics*, *47*, 1965, p. 118–120.

Kakutani, S., 'A generalization of Brouwer's fixed point theorem', *Duke Mathematical Journal*, *8*, 1941, p. 457–459.

Kantorovich, L.V., 'Mathematical methods for the organization and planning of production', 1939; translation in: *Management Science*, *6*, 1960, p. 366-422.

Kolm, S.-G., 'Décisions et concernements collectifs; contribution à l'analyse de quelques phénomènes fondamentaux de l'organisation des sociétés', *Analyse et Prévision*, 1967.

Kolm, S.-G., *Prix Publics Optimaux*, Centre National de la Recherche Scientifique, Paris, 1969.

Koopmans, J.G., 'De mogelijkheid van een meervoudig economisch evenwicht', *De Economist*, *81*, 1932, p. 679–702, 766–786, 841–856.

Koopmans, T.C., 'Analysis of Production as an Efficient Combination of Activities', in: *Activity Analysis of Production and Allocation* (ed. T.C. Koopmans), John Wiley, New York, 1951, p. 33-97.

Koopmans, T.C., *Three Essays on the State of Economic Science*, McGraw-Hill, New York, 1957.

Kornai, J., *Anti-Equilibrium*, North-Holland, Amsterdam, 1971.

Kornai, J., and T. Liptak, 'Two-Level Planning', *Econometrica*, *33*, 1965, p. 141–169.

Lange, O., 'On the Economic Theory of Socialism', *Review of Economic Studies*, *4*, 1936, p. 53–71 and 123–142.

Lange, O., *Political Economy*, Vol. I., 1959. Pergamon Press, London, 1963.

Lange, O., *Wholes and Parts*; *A General Theory of System Behavior*, (translation from Polish text, 1962), Pergamon Press, Oxford, 1965.

Lerner, A.P., *The Economics of Control*, Macmillan, New York, 1944.

Lindahl, E., *Die Gerechtigkeit der Besteuerung*, Lund, 1919, Chapter 4: 'Positive Lösung' translated as 'Just taxation – A positive solution' in: *Classics in the Theory of Public Finance* (eds. R.A. Musgrave and A.T. Peacock), Macmillan, London, 1967, p. 168–176.

Lipsey, R.G., and R.K. Lancaster, 'The general theory of second best', *Review of Economic Studies*', *24*, 1956, p. 11–32.

Makarov, V.L., and A.M. Rubinov, 'Superlinear point-set maps and models of economic dynamics', *Russian Mathematical Survey*, *25*, 1970, p. 125–169.

Malinvaud, E., 'Decentralized Procedures for Planning' in: *Activity Analysis in the*

Theory of Growth and Planning (eds. E. Malinvaud and M.O.L. Bacharach), Macmillan, London, 1967, p. 170–208.

Malinvaud, E., 'Notes sur l'étude des procédures de planification' *The Canadian Journal of Economics, 1*, 1960, p. 16–36.

Malinvaud, E., 'Risk-taking and Resource Allocation', in *Public Economics*, Proceedings of a I.E.A. Conference, (eds. J. Margolis and H. Guitton), Macmillan, London, 1969a, p. 222–246.

Malinvaud, E., *Leçons de théorie micro-conomique*, Dunod, Paris, 1969b. Translated as *Lectures on Micro-Economic Theory*, North-Holland, Amsterdam, 1972.

Malinvaud, E., 'A Planning Approach to the Public Good Problem', *Swedish Journal of Economics, 73*, 1971, p. 96–112.

Malinvaud, E., 'Prices for individual consumption, quantity indicators for collective consumption', *Review of Economic Studies, 39*, 1972, p. 385–405.

Marschak, J., 'Elements for a Theory of Teams', *Management Science, 1*, 1955, p. 127–137.

Marschak, J., and R. Radner, *Economic Theory of Teams*, Yale University Press, New Haven, 1972.

Marschak, T.A., 'Centralization and Decentralization in Economic Organizations', *Econometrica, 27*, 1959, p. 399–430.

Marschak, T.A., 'Computation in Organizations: Comparison of Price Mechanisms and Other Adjustment Processes', in *Risk and Uncertainty*, Proceedings of a I.E.A. conference, (eds. K. Borch and J. Mossin), Macmillan, London, 1968, p. 311–356.

Marx, K., *Das Kapital. Kritik der politischen Ökonomie*, Band 1, Hamburg, 1871.

McKenzie, L.W., 'On equilibrium in Graham's model of world trade and other competitive systems', *Econometrica, 22*, 1954, p. 147–161.

Meadows, D.L., *The Limits To Growth*, A Report for the Club of Rome Project on The Predicament of Mankind, Universe Bootes, New York, 1972.

Milleron, J.-C., 'Duality in consumer behavior analysis', paper presented at the European Meeting on Statistics, Econometrics and Management Science, Amsterdam, Mimeographed I.N.S.E.E., Paris, 1968.

Milleron, J.-C., 'L'approche par dualité dans la théorie des effets externes et des consommations collectives', paper presented at the Winter Meeting of the Econometric Society, Blaricum, 1969.

Milleron, J.-C., 'Theory of Value with Public Goods: A Survey Article', *Journal of Economic Theory, 5*, 1972, p. 419–477.

Mirrlees, J.A., 'An exploration in the theory of optimum income taxation', *The Review of Economic Studies, 38*, 1971, p. 175–208.

Mirrlees, J.A., 'On producer taxation', *The Review of Economic Studies, 39*, 1972, p. 105–111.

Morgenstern, O., 'Thirteen critical points in contemporary economic theory: an interpretation', *The Journal of Economic Literature, 10*, 1972, p. 1163–1189.

Muench, T.J., 'The core and the Lindahl equilibrium of an economy with a public good: an example', *Journal of Economic Theory, 4*, 1972, p. 241–255.

Musgrave, R.A., 'A multiple theory of budget determination', *Finanzarchiv*, 1957, p. 333–356.

Musgrave, R.A., *The Theory of Public Finance*, A Study in Public Economy, McGraw-Hill, New York, 1959.

Musgrave, R.A., 'Provision for Social Goods' in: *Public Economics*, Proceedings of a I.E.A. Conference, (eds. J. Margolis and H. Guitton), Macmillan, London, 1969, p. 124–144.

Musgrave, R.A., and A.T. Peacock, eds., *Classics in the Theory of Public Finance*, Macmillan, London, 1967.

Nauta, D., *Logica en model*, De Haan, Bussum, 1970.

Negishi, T., 'The Stability of a Competitive Economy: A Survey Article', *Econometrica*, *30*, 1962, p. 635–669.

Nikaido, H., *Convex Structures and Economic Theory*, Academic Press, New York, 1968.

Pareto, V., *Manuel d'Economie Politique*, Marcel Giard, Paris, 1927.

Pierson, N.G., 'Het waardeprobleem in een socialistische maatschappij', *De Economist*, *51*, 1902, p. 421–456; translated as: 'The Problem of Value in the Socialist Community', in *Collectivist Economic Planning* (ed. F.A. von Hayek); Routledge, London, 1935, p. 41–85.

Pikkemaat, G.F.W.M., *Informatie, Communicatie en Economische Organisatie*, Proefschrift, Groningen, 1969.

Roberts, D.J., 'Existence of Lindahl equilibrium with a measure space of consumers', Discussion paper 1, Northwestern University, Evanston, Illinois, 1972.

Robinson, J., *Economic Philosophy*, Double day, New York, 1964.

Rockafellar, R.T., 'Monotone processes of convex and concave type', *Memoirs of the American Mathematical Society*, Number 77, American Mathematical Society, Providence, R.I., 1967.

Rockafellar, R.T., *Convex Analysis*, Princeton University Press, Princeton, N.J., 1970.

Rockafellar, R.T., 'Convex algebra and duality in dynamic models of production', Preprint no. 8 of the Symposium on Mathematical Methods of Economics, Polish Academy of Sciences, Warsaw, 1972.

Rousseau, J.J., *Du contrat social, ou principe du droit politique*, 1762.

Roy, R., *De l'utilité, contribution à la théorie des choix*, Hermann, Paris, 1942.

Ruys, P.H.M., 'Prijs en vooruitgang', Paper submitted for the drs. degree to Prof. D.B.J. Schouten, Katholieke Hogeschool, Tilburg, 1963.

Ruys, P.H.M., 'Procedure to determine a program in an economy where only collective goods are available', Paper presented at the Second World Congress of the Econometric Society, Cambridge; mimeographed Katholieke Hogeschool, Tilburg, 1970.

Ruys, P.H.M., 'A procedure for an economy with collective goods only', Research memorandum 20, Tilburg Institute of Economics, 1971a.

Ruys, P.H.M., 'Centralized and decentralized resource allocation processes: a survey', Paper presented at the European Research Workshop in Economic Theory, Bergen, 1971 (sponsored by the International Economic Association); mimeographed Katholieke Hogeschool, Tilburg, 1971b.

Ruys, P.H.M., 'The relation between dual linear programs and their programs in the dual spaces', Paper presented at the European Meeting of the Econometric Society, Budapest; mimeographed Katholieke Hogeschool, Tilburg, 1972a.

Ruys, P.H.M., 'On the existence of an equilibrium for an economy with public goods only', *Zeitschrift für Nationalökonomie*, *32*, 1972b, 189–202.

Ruys, P.H.M., 'Production correspondences and convex algebra' in: *Seminar on Production Theory* (eds. W. Eichhorn, R. Henn, O.O-pitz and R.W. Shephard), Lecture Notes in Economics and Mathematical Systems, Springer-Verlag, Berlin, 1974.

Samuelson, P.A., 'A note on the pure theory of consumer's behavior' *Economica*, N.S., *5*, 1938, p. 61–71 and 353–354.

Samuelson, P.A., 'The Pure Theory of Public Expenditures', *The Review of Economics and Statistics*, *36*, 1954, p. 387–389.

Samuelson, P.A., 'Diagrammatic Exposition of a Theory of Public Expenditure', *The Review of Economics and Statistics*, *37*, 1955, p. 350–356.

Samuelson, P.A., 'Aspects of public expenditure theories, '*The Review of Economics and Statistics*, *40*, 1958, p. 332–338.

Samuelson, P.A., 'Pure Theory of Public Expenditure and Taxation', in *Public Economics*, Proceedings of a I.E.A. conference (eds. J. Margolis and H. Guitton). Macmillan, London, 1969, p. 98–123.

Sawyer, W.W., *Wegwijs in de wiskunde*, Het Spectrum, Utrecht, 1962.

Sax, E., *Grundlegung der theoretischen Staatswirtschaft*, Vienna, 1887.

Scarf, H., 'An analysis of markets with a large number of participants' in: *Recent Advances in Game Theory*, The Princeton University Conference, Princeton, 1962.

Schmeidler, D., 'Competitive equilibria in markets with a continuum of traders and incomplete preferences', *Econometrica*, *37*, 1969, p. 578–585.

Schumpeter, J., *Theorie der wirtschaftlichen Entwicklung*. Eine Untersuchung über Unternehmergewinn, Kapital, Kredit, Zins und den Konjunkturzyklus, 1913. Duncker und Humblot, München, 1926.

Schumpeter, J.A., *Capitalism, Socialism and Democracy*, Allen and Unwin, 1943.

Sen, K., *Collective Choice and Social Welfare*, Oliver & Boyd, Edinburgh, 1970.

Shannon, Cl. E., and W. Weaver, *The Mathematical Theory of Communication*, the University of Illinois Press, Urbana, 1949.

Shephard, R.A., *Cost and Production Functions*, Princeton University Press, Princeton, 1953.

Shephard, R.W., *Theory of Cost and Production Functions*, Princeton University Press, Princeton, New Jersey, 1970.

Shephard, R.A., 'Semi-homogeneous production functions and scaling of production', College of Engineering, University of California, Berkeley, Cal., 1972.

Shubik, M., 'Edgeworth market games', in *Contributions to the Theory of Games IV* (eds. A.W. Tucker and R.D. Luce), Princeton University Press, Princeton, 1959, p. 267–278.

Shubik, M., 'On different methods for allocating resources', *Kyklos*, *23*, 1970, p. 332–337.

Smith, A., *An inquiry into the nature and causes of the wealth of nations*, London, J.M. Dent, 1960.

Smithson, R.E., 'Multifunctions', *Nieuw Archief voor Wiskunde* (3), *XX*, 1972, p. 31–53.

Starrett, D.A., 'A note on externalities and the core', *Econometrica*, *41*, 1973, p. 179–183.

Stevers, Th. A., 'Een economische analyse van het democratisch proces', 1967. *Maandschrift Economie*, Lustrum-nummer, 1968, p. 37–70.

Taylor, F.M., 'The Guidance of Production in a Socialist State', *American Economic Review*, *19*, 1929, p. 1–8.

Tinbergen, J., *Centralization and Decentralization in Economie Policy*, North-Holland, Amsterdam, 1954.

Tinbergen, J., 'De optimale organisatie der economische beslissingen', *Mededelingen der Koninklijke Nederlandse Akademie van Wetenschappen, afdeling letterkunde, 24* 7, 1961.

Tinbergen, J., *Central Planning*, Yale University Press, New Haven, 1964.

Uzawa, H., 'Duality principles in the theory of cost and production', *International Economic Review*, *5*, 1964, p. 216–220.

Verheyen, P.A., 'De input-output-analyse binnen de onderneming', Intree-rede, Winants, Heerlen, 1965.

Vind, K., 'Edgeworth-allocations in an exchange economy with many traders', *International Economic Review*, *5*, 1964, p. 165–177.

Von Mises, L., 'Die Wirtschaftsrechnung im sozialistischen Gemeinwesen', *Archiv für Sozialwissenschaften und Sozialpolitik*, *47*, 1920; translated as: 'Economic Calculation in the Socialist Commonwealth', in *Collectivist Economic Planning* (ed. F.A. von Hayek), Routledge, London, 1935, p. 87–130.

Von Neumann, J., 'Über ein ökonomisches Gleichungssystem und eine Verallgemeinerung des Brouwerschen Fixpunkt Satzes', *Ergebnisse eines Mathematischen Kolloquiums*, *8*, Franz Deutieke, Leipzig, 1937. Translated as 'A model of general economic equilibrium', *Review of Economic Studies*, *13*, 1945 p. 1–9.

Wald, A., 'Über einige Gleichungssysteme der Mathematischen Ökonomie', *Zeitschrift für Nationalökonomie*, *7*, 1936, p. 637–670. Translated as 'On some systems of equations of mathematical economics', *Econometrica*, *19*, 1951, p. 368–403.

Walras, L., *Eléments d'économie politique pure*, L. Corbaz, Lausanne, 1874. English translation of the definitive edition by William Jaffé, *Elements of Pure Economics*. Allen and Unwin, London, 1954.

Weddepohl, H.N., *Axiomatic Choice Models and Duality*, Rotterdam University Press, Rotterdam, 1970.

Weddepohl, H.N., 'Duality and equilibrium', *Zeitschrift für Nationalökonomie*, *32*, 1972, p. 163–187.

Weddepohl, H.N., 'Dual sets and dual correspondences and their application to equilibrium theory', Research memorandum 38, Tilburg Institute of Economics, Tilburg, 1973a.

Weddepohl, H.N., 'Duals of integrals of correspondences, applied to a market', CORE discussion paper 7345, Leuven, 1973b.

Weitzman, M., 'Iterative Multilevel Planning with Production Targets', *Econometrica*, *38*, 1970, p. 50–65.

Younes, Y., 'Indices prospectifs quantitatifs et procédures décentralisées d'élaboration des plans', *Econometrica*, *40*, 1972, p. 137–146.

Index

INDEX OF AUTHORS

INDEX OF SUBJECTS

B 1565 Printed in Belgium by Ceuterick s.a.
Brusselse straat 153 B 3000-Louvain
A. Struyf Oude Baan 353 B 3040-Korbeek-Lo